Sunday, September 9, 2013

Today I had one of those *a-ha* moments that make me say that reason - and whatever intelligence we may be endowed with - gives us some of life's most refined and intense pleasures.

Concepts, words and ideas reorganized themselves in my mind in a slightly different way and, seemingly out of thin air, I conceived of a little phrase. It is just a short sentence, but I immediately sensed I was on to something big.

These few words in fact resume, encapsulate, condense and express all that I have learned through my studies and counseling practice. They are immediately understandable - hence their real strength - but they also are highly symbolic. They lend themselves to being unpacked, explored, and meditated on to incredible depths.

What came to my mind is the following:

We are not physical bodies,
with a consciousness that we lose at death.

We are consciousness,
with a physical body that we lose at death.

Adventures in Psychical Research

A medical doctor's exploration
of the nature of consciousness
and its survival to bodily death

Piero Calvi-Parisetti, MD

OpenMind

OpenMind Publishing – January 2015

ISBN-13: 978-1506147031

ISBN-10: 1506147038

© Piero Calvi-Parisetti

PIERO CALVI-PARISETTI, M. D.

 piero@drparisetti.com

 www.drparisetti.com

 facebook.com/parisettiDR

 @drParisetti

Contents

Foreword	1
Introduction	5
Between a rock and a hard place	7
52 per cent of people "overjoyed" to see a ghost	9
Consulting Spirit	11
Life after life	13
Finally, a theory for PSI	15
The problem with reincarnation	17
When religion doesn't help	21
The "other side": real help for grief (I)	25
The "other side": real help for grief (II)	29
The "other side": real help for grief (III)	35
The "other side": real help for grief (IV)	39
The "other side": real help for grief (V)	43
The powerful illusion (I)	47
The powerful illusion (II)	51
Proof of hell?	53
"ATTEND"ing to traumatic grief	57
Do mediums really talk to the dead? (I)	61
Do mediums really talk to the dead? (II)	65

Do mediums really talk to the dead? (III)	69
Do mediums really talk to the dead? (IV)	73
Afterlife science in support of the bereaved	77
Guess what – paranormal raps are… paranormal!	79
Mind is not in the brain	83
Assisted After-Death Communication (I)	87
Assisted After-Death Communication (II)	91
Mind and Molecules	95
What's It Like on the Other Side?	99
A Case for Blind Rage	103
An honors degree in mathematics with no detectable brain?	107
Do I believe in life after life?	109
The ghost in the machine	111
Is there any truth in the law of attraction?	115
Awareness during general anaesthesia	119
Reasons for wanting to know	123
Pseudoscience	127
PSI – Are We "Cherry picking"?	131
Are we conscious despite our brain? (I)	135
Are we conscious despite our brain? (II)	137
Are we conscious despite our brain? (III)	143
The science of mediumship (I)	145
The science of mediumship (II)	151
The science of mediumship (III)	155
The science of mediumship (IV)	159
The science of mediumship (V)	163
Amongst mediums	167
Spirit and the process of grief recovery	171
Being a medium (I)	177

Being a medium (II)	181
Being a medium (III)	185
Being a medium (IV)	189
The Departed Among the Living	193
What I believe (I)	195
What I believe (II)	199
Rationalist Spirituality	203
An extraordinary night of mediumship	207
Another case for blind rage	211
Awareness during resuscitation	217

Foreword

> *We are not physical bodies,*
> *with a consciousness that we lose at death.*
> *We are consciousness,*
> *with a physical body that we lose at death.*

As I opened the manuscript Dr Parisetti had sent me and found this beautiful and profound statement on the very first page, I felt moved and I immediately wanted to be the first person to officially quote it in writing. I feel it expresses the essence of who (and what) we are. If you, the reader, will remember any one thing from this precious book, I hope (and trust) it will be those twenty-five words.

Apart from this, you are about to read a delicious book. A book you can savor over many meals, or experience as tasty snacks whenever you wish. If you are like me, you will revisit particular essays as the mood moves you. In fact, I can't recall the last time I had so much enjoyment (and fun) reading a serious book about psychical science (or any science, for that matter).

Yes, because, on top of being delicious, this is a serious book indeed. It is a collection of 57 articles which had originally appeared as blog postings on Dr Parisetti's website. In the introduction, the author refers to the view held by some critics that "blogging is to literature what graffiti is to visual arts." In an amazing coincidence, immediately before I received the manuscript, my wife Rhonda – who has a bachelors and master's degree in painting – and I watched a movie which I had never heard of titled "Exit through the Gift Shop" – a documentary all about serious graffiti artists. Watching that documentary and, immediately after, reading the exceptional articles in this collection made me reflect on the real meaning of words like "art" and "literature".

In his writing, Piero – whom I have the pleasure of knowing personally – often emphasizes the importance of "open-minded recognition" and I wish to highlight his open-minded adventurous approach to psychical research. One cannot underestimate the importance of encouraging open-mindedness, especially about controversial topics. Having spent thirty years myself investigating frontier subjects, I am glad to see signs of increased interest and openness on the part of the larger scientific and academic community. Based on an international summit I co-organized on "post-materialist science, spirituality, and society," co-sponsored by the University of Arizona and Columbia University and hosted by Canyon Ranch in Tucson, we created a special forum (www.opensciences.org) to foster exploration and understanding about these topics. We chose the words "open sciences" very carefully. In his writing and scholarly work, Piero epitomizes this approach to science and life.

What do I appreciate and admire most about this book? The truth is that it is difficult to choose, and after serious reflection, I decided not to try to choose. Dealing with a range of diverse but closely related subjects, including the process of grief, life after death, the relationship between mind and brain, psi research, or what it is like to be a medium, the articles are all presented and written with a clarity and aliveness that is both enlightening and enlivening. In fact, I find it hard to believe that Piero is trained as a medical doctor. He writes too well!

One of the things I especially appreciate is that, although he clearly knows how to write, and chooses his words carefully, he is not afraid to pull punches and tell it as he sees it. When reading, for instance, statements like "This, the dogma of materialism, is the real pseudoscience. We are constantly lied to. We are denied the right to know. We are not allowed to get the full picture and to make our own judgment" I feel the words could have been mine, or anybody else's who has seriously investigated human consciousness and its possible survival of physical death. This feeling was immediately confirmed in another curious synchronicity. Three days before I sat down to read Piero's book and write this brief Foreword, I was interviewed by Roberta Grimes, an attorney by training, for a new radio show she is hosting in 2015. Roberta has published a number of books including *The Fun of Dying* and *The Fun of Staying in Touch*. I mention this particular radio interview because Roberta began and ended it with poignant comments about how conventional science is lying to us about these topics and that we are being denied access to the greater truth. I rarely hear interviewers speak so forcefully about these issues.

Finally, it is evident to me that not only Piero really cares about these topics, but he deeply cares about what we, his readers, make of all this. In fact, his blog serves as a platform for a continuous exchange of ideas between him and his readers, and – very interestingly – amongst the readers themselves. I therefore encourage you, if you are so moved, to join this virtual community of people with similar interests and share with Piero and others your thoughts and feelings about this book. At the same time, I would like to wholeheartedly encourage Piero to keep writing these stimulating and insightful blogs and hopefully share them with us in future volumes.

Gary E. Schwartz PhD

Professor of Psychology, Medicine, Neurology, Psychiatry, and Surgery, and Director of the Laboratory for Advances in Consciousness and Health, at the University of Arizona.

Introduction

I once heard a self-appointed literary critic affirming that "blogging is to literature what graffiti is to visual arts".

There are two ways, I think, of looking at this statement. You can take it as derogatory and self-righteous, belittling both blogging and graffiti by comparing them to implicitly superior forms of expression. Or you can take it as an open-minded recognition that both literature and the visual arts are diverse, evolving fields, and new languages and modalities have emerged in the recent past as products of the world we live in.

The reason I am talking about this in the introduction is that this book is a collection of the blog postings I have published between 2013 and 2014 on my website, drparisetti.com, and I shamelessly maintain that this is indeed literature. My conviction rests on a few key arguments.

Firstly, with these writings I had the explicit intent of expressing myself, of saying something, of sharing original ideas and lesser-known information. These, in short, were not compositions thrown together and published – as is the case for certain blogs – to increase the number of *hits* on a website or to get a better ranking on search engines.

My blog – and this book – are the outcome of my work as a scholar and a practitioner of applied psychical research. Psychical research and parapsychology (the terms are used interchangeably) are concerned with the scientific investigation of the ways that organisms communicate and interact with each other and with the environment that appear to be inexplicable within current scientific models. They are also concerned with the scientific investigation of phenomena suggestive of survival of personality of bodily death. *Applied* psychical research is a discipline that looks at practical applications of the body of knowledge acquired through scientific investigation.

As a medical doctor and a trained psychotherapist, my interest lies in the use of the results of psychical research to ease the suffering of two categories of people – those who are in pain over the loss of a loved one, and those who are distressed because of imminent death (their own or a loved one's).

And here comes the second reason that makes me believe that my blog postings are a form literature. The 57 articles that form this book are individual, stand-alone pieces that can be read in isolation. However, they also belong to a greater, coherent and consistent scheme and logical framework. Psychical research provides compelling evidence for the facts that mind is related to but independent from the physical brain and significant aspects of human personality survive the death of the body. These subjects – the mind/brain relationship and the survival hypothesis – provide the substance for most of my writings, but they are not dealt with as a matter of simple intellectual curiosity. The underlying angle, the common thread linking all of this work, is that these subjects are crucially important for the bereaved and the dying. When it is understood – rationally understood – that death does not equate with disappearance/annihilation, a considerable part of the fear of death and some of the pain of bereavement can be avoided.

Thirdly, and lastly, there is consistency of format, style and language. The vast majority of the articles take relevant – and often lesser known – pieces of information produced by psychical research, present them in a language adapted for a non-specialist public, put them into context as described above and offer my own, original reflections. Such consistency, I believe, makes them "natural" material for a book.

It is my hope that those who follow my weekly postings – and have contributed a large number of most interesting comments – will find it useful to have nearly two years of articles gathered into a single publication. And, especially, I hope that through this book my ideas will reach a larger public and inspire further interest and reflection.

Glasgow, December 2014.

Between a rock and a hard place

My job is not easy. Working with the bereaved and the dying can be very satisfying – if your aim is to be of help to others – but inevitably it exposes you to some humanly difficult situations.

However, what I find at times really challenging, as I follow my uncompromising, evidence-based, life after life approach, is the fierce opposition I encounter from two categories of believers. I find myself stuck between a rock and a hard place.

On the one hand, there are people of religious faith, for whom I have the greatest respect. They won't accept the conclusions I draw from evidence – and the direct experience of millions of fellow human beings – because they think these may contradict the teachings of their religion.

On the other, there are people of materialistic faith, unfortunately predominant in the scientific world, for whom I have considerably less respect. Their refusal is based on the belief – yes, it is a belief – that all that exists is matter. For them, human consciousness is either simply a product of the physical brain or – for the real extremists – it doesn't even exist – it's just an illusion! In any case, they say, when the brain stops working, anything we may call mind, or consciousness, comes to an end.

But the empirical evidence is there. It does not go away simply because some people don't like it. And, in terms of human experiences, the only way to refute the testimony of the thousands who, every day and all over the world, have near-death experiences and deathbed visions or experience communication with deceased loved ones, is to bury one's head in the sand and simply refuse to see.

Both empirical evidence and the experience of millions point to the fact that our mind, our consciousness, our personality are independent from the physical brain and go on existing after the death of the body.

Why not look at this evidence and these human experiences? Why not think for ourselves, draw our own conclusions, to the best of our knowledge and understanding? Then, a belief based on reason, and not – or not simply – on faith, can emerge, with extraordinary psychological benefits for those who are suffering.

52 per cent of people "overjoyed" to see a ghost

According to a survey by the Forever Family Foundation, 52% of the interviewees would be "overjoyed" and would "try to interact" if they were to see a deceased loved one while they were awake.

It would be easy to dismiss this as just a nice thought – the understandable desire on the part of a bereaved person to renew contact with somebody who's passed away. In fact, things are not so simple.

First of all, research tells us that some sort of interaction with the deceased is a very frequent phenomenon. According to the University of Chicago, for instance, as many as 43% of American adults report having had some sort of contact with a deceased loved one. And nearly one in three of the Americans who do not believe in life after life still report an interaction.

Many of these experiences can be explained, but some actually cannot. These include, for instance, cases in which there is sustained two-way communication, touch, and especially, the same apparition being seen by many witnesses at the same time. Experiences of interaction with the dead have been investigated by top researchers and in great detail. Whoever has bothered looking at such investigations understands that, collectively, the phenomenon cannot be explained away as fantasies, hallucinations, or misperceptions.

What is most important, however, is that these experiences are often very comforting for those who have them. They can significantly ease the pain of a loss, providing meaningful evidence that the loved one goes on living in a non-material dimension of existence.

This being said, it is also true that many people who have had such an experience find it difficult to accept it as reality. I have received many letters from persons who've had "signs" from a deceased loved one and found them comforting, but still ask me to confirm or validate their experience. And, forget talking about such things with your doctor, or with a traditional grief counselor! There is total rejection of life after life experiences from a large part of health professionals.

For this reason, critically examining the large body of evidence for life after life can be very helpful. It can relieve the suffering of grief, supporting a belief in the afterlife which is based on reason rather than on confused and unsubstantiated religious teachings. And, for those who've experienced an interaction with a discarnate personality, it provides the confirmation that this was indeed a real experience.

Consulting Spirit

I keep hearing from bereaved persons and their relatives about the near-total closure of the medical and counselling establishment regarding the possibility of life after life. Recently, one person was telling me that her counselor insisted that she should conform – in her grieving process – to the Elizabeth Kubler-Ross' "five stages of grief", a very popular model despite being unsupported by clinical evidence. When this person insisted on her right to grieve in her own way – and in particular alluded to the existence of an afterlife – she was basically told she needed medical attention.

It is therefore easy to fall into the generalization that the medical establishment in its entirety follows the materialist dogma. This, fortunately, is not true. Cracks are indeed opening and more and more physicians are "coming out" and speaking openly about the possibility of life after life.

Owing to the recent media hype about his book *Proof of Heaven*, we all know about academic neurosurgeon Eben Alexander and his life-changing near-death experience. But here I want to talk about somebody considerably less known and talked about, who has had the courage to speak openly about his own gift of mediumship and how he resorted to using this in his clinical practice.

When he wrote his book *Consulting Spirit: A Doctor's Experience with Practical Mediumship,* British physician Dr Ian Rubenstein had been a doctor for 34 years and a family physician for 28. It does not get any more mainstream/middle of the road than this in the medical profession.

Dr Rubenstein says "I don't come from a religious background at all. I'm from a non-practising, left-wing, Jewish background. All my family was, you could say, very anti-religious. I'm not a religious guy. Spirituality is not new to me. I'm as affected by new-age stuff as much as everybody else, but it's not native to my culture and background, and certainly not to my education."

He therefore shares his own astonishment when one day, in 2004, an Irish patient named Lucy came into his consultation room and "was inconsolable to the point where I did something I don't normally do. After 15 minutes of her crying and not getting any sense out of her, I actually in desperation printed a prescription for an antidepressant, which I handed to her.

As I was about to hand it to her, I felt this blow to the back of my neck and heard a voice behind me saying, "Ask her about her father." With that, over her left shoulder, I saw a mist. It was like I could see it not quite with my eyes, but it was almost like I could see it. It was more than just imagination or an image in my head. I could see this misty form of a guy which I could describe over her left shoulder.

I just said to her, "Lucy, tell me about your father." She looked at me, stopped crying, and said, "He was killed 38 years ago on the 8th of December. In two days' time would be the anniversary of his death. Do you think that's why I'm depressed?"

I said, "Did he look like...?" and I described the guy over her left shoulder. She said, "Yes, how did you know?" I said, "Lucy, I think I can see him over your left shoulder." She grabbed my arm, stopped crying, looked at me and said, "Doctor, you don't know how much this means to me. Thank you, so much."

In his honest, personal, captivating book, Dr Rubenstein recounts how he had to slowly come to terms with this and many more happenings that followed. And of how, at some stage, he decided to take the plunge and fully develop his suddenly-revealed mediumship capabilities.

What I particularly liked is that Dr Rubenstein makes no mystery of the fact that, once he had accepted that he was indeed a medium and he had fully developed his capabilities, he integrated this very unorthodox approach with his very orthodox practice of family medicine, with very positive outcomes for his patients.

Life after life

My Facebook friend Jim Beichler is a scientist. In a recent exchange we had, he made an interesting and very pertinent observation on the common use we make of certain terms or expressions. In a nutshell, he said that speaking about "life after life" and "life after death" actually makes no sense. He, himself convinced of the survival of human personality of bodily death, said that this confusion in semantics is actually counter-productive to our efforts to have an informed dialogue with the largest possible audience.

This observation had me thinking a lot. I can actually see Jim's point quite clearly. If we understand "life" as the functioning of the complex biological system that is our body, then it is clear that life comes to an end when this functioning stops. If we take this view, there can be neither life after life, nor life after death.

This, then, inevitably begs a more fundamental question: What is life? Without being a philosopher, I am very attracted by the generalities I know about the theory known as monistic idealism. This theory holds that consciousness, not matter, is the ground of all being. It is monist because it maintains that there is only one type of thing in the universe and idealist because it maintains that that one thing is consciousness. The 20th-century British scientist Sir James Jeans famously wrote that "the Universe begins to look more like a great thought than like a great machine".

The idea that consciousness is the ground of being is compatible with the masses of compelling evidence we have for both psychic powers and for the survival of human personality. Materialism is clearly not.

Therefore, human life is for me the activity of the mind rather than the activity of the body. What we call "ourselves", truly, is our thoughts, our memories, our awareness, our feelings – in one word, our mind.

But then I realize that, even with this idealist view of life, Jim's observation holds true. The expression "life after life" is a misnomer, because life (mind, consciousness) is not related to time as we understand it. There is no "before" or "after" – there is just "life". And, the expression "life after death" is also a misnomer, simply because there is no death.

And still, these expressions are very useful in my grief counselling work. They effectively communicate the basic message that life continues after the death of the physical body. In the midst of their suffering, the bereaved and the dying don't have much appetite for philosophy or semantics, and I am not sure that I am ready to abandon these trusted communication tools.

Finally, a theory for PSI

From the beginning of the controversy between proponents and skeptics of parapsychology, one of the major stumbling blocks has been the lack of a scientific theory which would account for the experimental data and would enable researchers to make predictions.

With much less hype and visibility than it would have deserved, such a theory was proposed in 2012 by North Carolina University Professor James C. Carpenter in one of the most remarkable books I've ever read.

First Sight is admittedly not an easy read. The professor is obviously incredibly learned and has put some stupendously sophisticated, solid and compelling thinking into it. However, communication is not necessarily his forte, and even the scientifically and philosophically minded reader has to put some effort into the more than 400 pages of heavy substance. I am just waiting for somebody to bring some of this brilliant thinking into the language and form of popular science.

The basic tenets of the theory are that, whilst often seen as supernatural, unpredictable and possibly dangerous, psychic activities are actually happening all the time and help us make sense of everyday experiences. Hence the title of the book, and the name of the theory itself – *First Sight*. This indicates that such faculties and activities are not secondary to "normal" psychological and conscious processes but in fact precede them and, largely, shape them.

According to *First Sight*, PSI is essential to the construction of experience. Extrasensory perception is the leading edge of the mind's ability to move to the next experience. Psychokinesis is the leading edge of the mind's ability *to move the next effect to its intention*. These PSI processes are continuously active but normally unconscious and implicit. This implies that all experience and the actualization of all intention begin at the PSI level of functioning. PSI is not "second sight" but "first sight".

Extrasensory perception is therefore proposed as an ongoing, continuous process which influences not only the way we move from one thought to another, but also how we perceive the world, how we react to stimuli, what we fear and what we don't, what we remember and what we don't, and, more in general, our entire personality. Only occasionally PSI functioning becomes conscious. Otherwise it remains below the level of consciousness as a very useful adaptive mechanism that helps us survive in a challenging environment.

The strength of the theory lies in the fact that it is based on evidence. It is not just some fancy speculation or smart fantasizing. Carpenter draws on a vast array of studies in contemporary psychology on problems as diverse as memory, perception, personality, creativity and fear. As I said, the theory brilliantly accounts for the research findings and provides the basis for a testable hypothesis.

Given the extraordinary importance of this development to the field of parapsychology, I am truly surprised that this is not all that is being talked about in psychical research circles. I am still to read any book review in the specialized press and I have only heard of a few of the leading thinkers referring to the theory and the book. Hopefully, that will come soon.

Otherwise, I'm not surprised that none of the many vociferous skeptics has as much as made a mention about all of this. Now, they will not be able to go on whining about the lack of a testable theory. The theory is there – it's solid, it's compelling and it supports and explains the mass of empirical evidence that's been gathered for PSI phenomena.

I'm sorry, Mr Skeptic, you now really seem to be stuck between a rock and a hard place.

The problem with reincarnation

A gentle warning: this article is not for "beginners". If you are a bereaved or a dying person and are interested in my grief counselling approach based on a rational belief in life after life, I suggest that you first familiarize yourself by looking at the other sections of the website. This blog post discusses more speculative aspects of the survival hypothesis and may actually result in confusion for those who are not familiar with the mass of compelling empirical and scientific evidence pointing to the existence of an afterlife.

On the other hand, if you are "in the know" you may have asked yourself a puzzling question. We have very strong evidence that the entities who communicate with us – through mediums and other means – from the nonmaterial dimension of life we call the spirit world (for lack of a better term) are indeed who they say they are: discarnate personalities who've had an earthly life and have now moved on to a different plane of existence. One may therefore conclude that humans live a material life and, when the body dies, they continue to exist as personality/consciousness in the spirit world.

However, we also have very convincing evidence for the fact that we, as humans, reincarnate. The extensive research on children who remember a past life in stunning detail and on birth defects associated with a violent death in a remembered previous life leave little doubt that reincarnation does occur. Furthermore, thousands of well researched cases of memories of previous lives recalled by patients under hypnotic regression further corroborate this hypothesis.

These two facts – the existence of discarnate personalities and reincarnation – are in apparent contradiction, or at least pose some tricky "procedural" questions. Does everybody reincarnate? And, in this case, how can one continue to exist in spirit and reincarnate? Do some people reincarnate and others do not? And, in this case, what are the rules that govern this process?

Let me make one thing clear. The answers to these and many other questions remain, as far as I am concerned, resolutely in the domain of speculation. There are things about which I am, to the best of my intellectual honesty, sure, and others about which I like to speculate and would never imagine presenting as "the truth". The movements between the material and nonmaterial dimensions of life are an example of the latter.

There are two interesting hypotheses that may help us make some sense of all this. The simplest one is that the idea of time we have during our earthly life is one of the many persistent illusions we fall under. What for us is clearly a "before" and an "after" has no meaning in the timeless ultimate reality which is described to us by mystics and spirit communicators alike as an "everlasting present". This may well be the case, but then the passing of time, this persistent illusion of ours, is somewhat shared by those in the spirit world. Evidential communications coming through mediums, for instance, often refer to happenings in the sitters' lives, with a clear sense of "before" and "after".

The other hypothesis pushes the envelope a little further, and stretches our capacities of understanding. It is beautifully articulated in a book by Julia Assante, PhD, called *Exploring the Afterlife*. Dr Assante is both an academic social historian and a medium and speaks with the authority of a scholar and a direct experiencer at the same time. Whatever I would say here would not do justice to her thinking - I strongly suggest that you read her book and get a fresh, interesting perspective on many afterlife issues.

In a nutshell, Dr Assante introduces the concept of a "nonlocal oversoul". Wow! Not as difficult as it sounds. Yes, we do exist in the material world. Yes, we continue to exist in the nonmaterial world. And, yes, we do reincarnate. However, since in reality there is no time, all this happens at the same time. She speaks about "the notion that we, like the dead, are projections into a plane where we create with the illusions of time, place, and matter." And, consistent with what we hear from patients in hypnotic regression, there is both an afterlife and a "pre-life".

If this were true, a part of "me" would be sitting here typing away, another part of me would be in spirit, another part would be reincarnated, another would be waiting to be reborn. The same "oversoul", the same individual expression of the ultimate consciousness which – we like to think – forms the ground of being.

Fascinating matters for reflection and meditation. Thank you, Dr Assante!

When religion doesn't help

Ever since I have been studying the evidence for survival of human personality of physical death, the question of religious beliefs has been nagging me. This has turned into an outright discomfort since I have moved into the field counselling for the bereaved and the dying.

It is difficult for me to approach this subject, because the last thing I want to do is to come across as disrespectful for what are likely to be the most cherished, fundamental beliefs for so many people around the world. I insist in saying that, although I am not religious myself, I greatly respect religions and, especially, religious people.

Nevertheless, I cannot help but notice that many common religious beliefs about death and the afterlife are: a) in sharp contrast with masses of evidence and testimonies consistently coming to us from different lines of investigation, and, especially; b) quite unhelpful for a person who is facing death or grieving the loss of a loved one.

I feel passionately about this problem, since a few years ago I had the painful experience of accompanying my dearest friend in a three year dramatic battle with cancer. What was painful was certainly not being at my friend's side during those difficult times. Rather, I consider that an enriching experience. What I simply cannot stand is the idea that he died an anguished man. A committed Catholic, he was convinced that he was going to face judgment for sins he believed he had committed.

At that time I already knew that we have no evidence whatsoever for this kind of judgment. All the testimony we have, from a bewildering range of sources, consistently speaks of a life review instead, in which we are not "judged" by others, but rather we are helped to make sense of the life we lived, understanding ourselves what was good and what was less good. Especially, we have not the tiniest evidence in support of things like hell or eternal damnation. We appear to learn, sometimes painfully, and then to progress.

Similar in many ways is the case of a neighbour – a lovely lady in her early fifties, who looks more like in her early seventies. For years, she's been grieving the premature death of her husband, and this has taken a big toll on her mind and body. When I said that I knew things that perhaps could help relieve some of her grief, she listened to me politely for a while, but then she said that she could not entertain the idea of survival – and, especially – of after death communication because of her religious thinking.

In all honesty, this makes me mad. If religions, as Karl Marx famously said, are the people's opium, then they should make people feel good, or at least less bad. They should certainly not add to the load of suffering connected with death and dying.

Furthermore, there is the problem of confusion, contradictions, and lack of clarity. I was raised a bland Catholic by a moderately religious family. That's the reason why I focus on Christianity/Catholicism here – I do not single that particular religion out or make any value judgment. The fact is that I am pretty convinced that if you were to ask ten Catholic or Christian people for a description of what happens after they die, you would get ten different answers. Where do people go before resurrection? Does resurrection happen for everybody? Do bad people go to hell immediately, or after judgment day? Do you have to be first resurrected and then, if bad, go to hell? What happens if you are bad and repent? Where, in the sequence of events, does purgatory lie? Etcetera.

I suspect that many teachings of religion may look absurd, inconsistent, and incoherent, even to the faithful. Many others are partially known or poorly understood (ask churchgoers to give you an explanation of the exact meaning of many of the things said every Sunday by the priest during Holy Mass, and you're in for a surprise). And still, people cling to these beliefs with fervour. That's what they've been taught – understandable or not, reasonable or not, comforting or not – and that's what they believe.

With the help of a great article by Miles Edward Allen, let's focus for a moment on the belief that speaking with the dead is sinful.

The main support for this belief in the Old Testament is from the book known as Leviticus. The key verse says "Do not turn to mediums or wizards, do not seek them out to be defiled by them". There is another verse in Leviticus and one in Deuteronomy that clearly judge those who make a practice of talking to dead people: "A man or a woman who is a medium or a wizard shall be put to death;" and "There shall not be found among you any one who ... is a medium, for they are an abomination to the Lord".

There can be no doubt that these statements are in the Bible and that they distinctly prohibit consultations with the spirit world. If you accept Leviticus and Deuteronomy as the inerrant word of the Almighty, then you would be wise to avoid any contact with mediums. But, before you make such a decision, you might want to know what else you are signing up for.

Have you ever eaten a rare steak? Or a fatty hamburger? Have you ever trimmed your hair or beard? Did you ever get a tattoo; peak at your brother in the nude; fail to stand when an old man enters the room? Have you ever worn a shirt of cotton or polyester blend? Perhaps you have cursed a politician? According to the Old Testament, all these acts and many, many others are sins against the Lord and are condemned just as strongly as consulting a medium.

Finally, and this is even more difficult to accept, some of the teachings do not even allegedly come from God. For instance, the ultimate nature of Christ and other essential aspects of the doctrine were decided upon by vote amongst bishops, most importantly during the Council of Nicaea in 325.

This may now really seem like a tirade against religion, and Christianity in particular. It is not. It is a tirade against the unnecessary suffering deriving from uncritically accepting and believing some – at times poorly understood – religious teachings. Teachings in support of which there is no evidence whatsoever.

Evidence from psychical research points to an entirely different view about life and the afterlife. However, I am certainly not advocating for switching one belief for another. My strong advice is to consider the evidence, study it, reflect upon it. Use your intelligence, your reason, and draw your own conclusions from the data. Chances are, as medical research proves, that you will emerge with a clearer – and extraordinarily more comforting – understanding of death and the afterlife.

The "other side": real help for grief (I)

The suffering caused by the death of a loved one ranks amongst the most distressing of human experiences. It is also, sadly, one of the most common. Just as each person is unique, however, each loss is unique and each individual's reaction to loss is unique. Medical research shows that there is no "right" or "wrong" way to cope with grief and there are no "better" or "worse" persons in the face of such intense suffering.

Following a loss, most people experience what is defined as normal grief - a period of intense sorrow, numbness and even guilt or anger. Whilst for many people these feelings gradually ease, for some the feelings of loss are devastating and do not improve with the passage of time. This is called complicated grief - a debilitating condition which requires professional attention.

The problem is – what kind of professional attention?

With this article I will start a mini-series exploring in some depth the science behind my "no compromise" approach to grief counselling. By "no-compromise" I mean an approach which remains science- and evidence-based, but which openly admits to the existence of an afterlife. No beating around the bush, no politically correct attitudes. Just rattle the cage and tell the bereaved and the dying that death as we commonly think of it doesn't exist – not as a matter or faith or wishful thinking, but based on masses of compelling evidence.

In this mini-series, we will look at the work of others who have pursued other "non-traditional" approaches to helping people recover from grief. All of these approaches have three things in common: a) they were introduced or investigated by scientists, medical doctors or mental health professionals; b) they either imply or are based outright upon the existence of an afterlife; and c) they are extraordinarily effective.

The primary reason for looking at these non-traditional approaches is their effectiveness. Do we need alternative approaches to grief counselling? Aren't we happy with the established methods? Alas, no, we are most definitely not.

In this first chapter of the mini-series, I will attempt to answer the fundamental question I asked a moment ago – What kind of professional attention do the bereaved and the dying need?

Unresolved grief is considered pathological in stage models (e.g. Kübler-Ross, 1969) and the goal of traditional psychotherapy sessions is "working through the stages," and resolving the sense of loss. Already here, at the foundation level, as famously said by the Apollo 13 astronauts after their vessel was hit by a meteorite, "Houston, we have a problem…"

The problem is that the stage theory has received precious little experimental support. On the contrary, as I already pointed out in a previous post, research indicates that the stages of grief in fact do not exist. The very Dr Elizabeth Kubler-Ross, towards the end of her life, openly admitted that this was not a universally applicable model, and that she didn't think everybody had to go through the stages in sequence. Nevertheless, as the model is simple, logical and appealing, it spread like a cholera epidemic, "infecting" the entire grief counselling profession and expanding into other areas of psychology and sociology (the stages of grieving a job loss…).

What can we say? Well, if the traditional approaches worked, even if they were based on a shaky premise, who cares? But, alas, they don't.

In a 2007 paper (*What has become of grief counselling? An evaluation of the empirical foundations of the new pessimism* published in Professional Psychology: Research and Practice), researchers Larson and Hoyst pointed to the popular yet pessimistic consensus in the grief and bereavement literature that grief counselling was at best ineffective and at worst harmful to clients seeking help.

Similarly, in a 2008 meta-analysis of traditional psychotherapeutic grief treatment outcomes (*The effectiveness of psychotherapeutic interventions for bereaved persons: A comprehensive quantitative review* published in the Psychological Bullettin), Currier, Neimeyer, and Berman revealed a "discouraging picture for bereavement interventions" which added "little to no benefit beyond the participants' existing resources and the passage of time".

Traditional interventions are based on an unsupported theoretical model and do not seem to work. And still, this is what the overwhelming majority of grieving people are offered. This is what they get, what they pay for.

The good news, however, is that non-traditional interventions and experiences have been repeatedly demonstrated to dramatically diminish and even entirely alleviate grief. This is what we are going to look at in this mini-series. Scientific research – yes, the same "science" which too often ridicules our claims about the afterlife – shows that approaches implying the existence of an afterlife do work.

Let me conclude this initial article with a very telling quote, which perfectly captures and summarizes my entire approach to grief counselling. It doesn't come from a New Age guru or self-appointed spiritual master. It comes from Dr Carl Wickland, Director of the National Psychopatic Institute of Chicago, US.

> "What becomes of the Dead? This problem is of vital interest to the patient who lingers on the borderland of transition, doubtful of the future, or perhaps trembling in fear of his probable condition after the tomorrow of death. Should it not be the noblest part of the physician's calling, in such situations, to be in a position to assure his patient from actual knowledge, that there is no death, but a birth into new fields of activity and opportunities in the higher mental spheres?"

The "other side": real help for grief (II)

In this second chapter of my mini-series on non-traditional, "afterlife-based" approaches to grief counselling, we discuss the after-death communication technique developed by Dr Raymond Moody, in which and with whom I personally trained in 2012. This is a bit of a long post, but the subject is so fascinating that I think it is well worth the time you will devote to reading it. If you have not done so already, I strongly suggest that you read first the initial chapter of the mini-series.

An Oracle of the Dead for the 21st century?

In ancient Greece, those wanting to consult what we would call today a fortune teller didn't have to travel far from main urban centers. The famous Delphi Oracle, for instance, delivering famously ambiguous statements about the future, was conveniently located less than 100 kilometres from Athens. Those wanting to get in touch with the dead, on the contrary, had to undertake an arduous physical and psychological journey. They would travel for hundreds of kilometres in order to reach remote locations, were they would spend days in underground caves preparing for the momentous culmination of the process. This consisted in staring into cauldrons filled with oil. There, from the seemingly bottomless depth behind the surface of the oil, reflecting the darkness of the cave, dimly illuminated by a distant fire, the dead would appear, giving people the unshakeable conviction of having been reunited with departed loved ones and, we are told, relief from the suffering of grief.

Spontaneous apparitions of the dead are commonly reported still in our day. Medical reports show that at least half of all persons whose spouse dies report a spontaneous contact from that person after death. Similar reports often come from parents who have lost a child. This phenomenon is of great interest to those who work to alleviate the many negative consequences of unresolved grief. On the one hand, there is a grim consensus amongst researchers that traditional psychotherapeutic approaches are basically useless, at best, and at worst can actually worsen the symptoms of grief. On the other, research indicates that people who had an experience of after death communication with a deceased loved one, either spontaneously, through a medium or with other techniques, show marked and lasting improvements in their psychological well-being.

In the early 1990s, Dr Raymond Moody, a psychiatrist and then Professor of Psychology at West Georgia University was already a world celebrity for having been the first to publish research, in the mid-1970s, about the phenomenon of Near-Death Experiences He decided to experiment with the methods used by the Greeks some 2,500 years ago to see whether these beneficial experiences of reunion could be replicated outside the haphazard, unpredictable domain of spontaneous apparitions. The results he and other researchers obtained went way beyond expectations: a significant proportion of the subjects who undergwent this procedure are absolutely convinced that they had an experience of reunion with a deceased loved one. This methodology provides today a useful setting for research and, especially, has been demonstrated to provide a range of beneficial effects for those stricken by grief.

The procedure

The procedure runs for the best part of a day and includes an important preparation stage. In a talk-therapy setting, the patient is asked by the counsellor to reminisce about the deceased loved one, investigating memories, emotions, and revealing the details of the nature of the relationship. Very importantly, any "unfinished business" or unresolved issue the patient feels he or she has with the deceased person is discussed.

In the core part of the procedure, the patient sits comfortably for a couple of hours on a recliner chair in a completely darkened meditation room, called by Dr Moody "psychomanteum" in memory of the Oracle of the Dead in ancient Greece. Behind the chair, a dim light provides a modicum of illumination. From the recliner, all the patient has to do is to very calmly gaze into a large mirror, which is placed on a facing wall, in a slightly higher position, so that it does not reflect the image of the patient.

After the psychomanteum session, the patient and I take another good moment to discuss the experience and any resulting feelings.

The results

In his initial, seminal work (1992), Dr Moody reported that about half of the people who underwent his technique reported seeing apparitions of the dead in the mirror. Other researchers using the psychomanteum have had varying degrees of success in producing reunion experiences. Drs Dean Radin and Jannine Rebman in 1996 also reported that just over half of the subjects who spent time in their electronically-monitored psychomanteum had sensed the presence of a deceased person. Dianne Arcangel in 1997 had a strong degree of success, with 58 of her 68 participants (85%) having a reunion experience. Fifty-five of them reported a visual apparition as part of their experience. Dr Arthur Hastings and his associates at the Institute of Transpersonal Psychology in California had 27 people participate in their 2002 psychomanteum study. Of these, 13 (48%) described a reunion experience. In 2004, Dr William Roll held a series of psychomanteum workshops with 41 people who sought a reunion with deceased friends and relatives. In this case, only nine of the 41 people (22%) experienced a reunion.

On one point all the researchers agree: success rates in reunion experiences vary, and the nature of the reunion may be very different – from a "sense of presence" to voices, smells and touches, all the way up to fully-formed visual images of the deceased, talking and interacting as in real life – but most grieving people who go through a psychomanteum experience see their grief substantially reduced and a number of psychological health indicators substantially improved.

Critics claim that the encounters could be hallucinations and wish fulfilment of grieving relatives. Dr Hastings, amongst others, points out that whatever the origin of the contacts, a large body of medical evidence shows they are supportive and reassuring. In her PhD dissertation, for instance, Rebecca J. Mertz examined pre to post Likert-scale measurements with 100 participants in her psychomanteum study, demonstrating statistically significant reductions in bereavement components such as feelings of anger and guilt, as well as significant increases in positive feelings. The psychomanteum process was commonly described as helpful, healing, comforting, and peaceful. As well, it was frequently reported that the experience left the participant with a sense of well-being and a feeling of serenity and acceptance. Almost all of the participants reported subtle, embodied phenomena that engaged multiple dimensions of conscious awareness.

Similarly, in the concluding remarks of their own study, Dr Hastings and his collaborators state: "The statistical analyses indicate that there were strong shifts in unresolved feelings, according to the self-reports. Changes were in a direction of resolution, healing, and comfort. The impact of this process on persons' feelings and lives is a strong effect for a one time experience, and suggests that a Psychomanteum setting can have some use in encouraging grief reduction."

The experience

Now – this is what research tells us. But I believe I may have aroused your curiosity as to what the experience actually consist of. I see no better way than to let Dr. Moody himself describe, as he did in the groundbreaking article of 1992 in the scholarly Journal of Near-Death Studies, what the subjects of his experiment reported to him.

> "Never once during the planning stage of this experiment did it enter my mind that the apparitions might talk as well as appear; yet that is precisely what happened in six of the sixteen apparitional encounters. In these cases, complex quasi-communications were reported to take place between subjects and apparitions. These ranged from a few words of reassurance and love to lengthy, involved, interactive communication.
>
> A journalist in her early 60s, a well-respected editor of a large newspaper, came to the psychomanteum hoping to see her son who had committed suicide some months earlier. When he appeared, he told her that he was fine and that he loved her. During another session, a man in his mid-30s saw his deceased mother who, expressed surprise and curiosity as to how her son had come to where she was. One woman, a 44 year old counselor, described lengthy conversations with her deceased father.
>
> In several of the experiences in which no "verbal" exchanges took place, there was, nevertheless, communication of some other kind. One man reported that the apparition of his mother acknowledged his presence and conveyed to him nonverbally that she felt the woman to whom he had recently become engaged was good for him and that she approved. This man had been very close to his mother, and this reflected a concern that had been on his mind. Another man, a psychologist, saw two of his cousins; he felt that they

acknowledged his presence during the encounter.

Another surprise was that not every subject met the person he or she had expected to see. A businessman in his 70s spent a long day with me preparing to see his father; it was his departed business partner who showed up instead. A woman tried to see her husband but it was her father who came to her.

Subjects so far have been unanimous in asserting that what took place was completely real. When I asked a smiling 55 year old ophthalmologist what his judgment was, whether he felt his experience was a projection of his own mind, he said with emphasis, "Oh, no! I saw my mother!" A woman assured me with complete confidence and equanimity that she had indeed been with her deceased father. A journalist insisted that he had in fact seen and visited with his father during his session in the mirror booth.

Thirteen of the subjects described a compelling sense of the presence of their lost loved ones during these encounters, a feeling they voiced in such phrases as "I know he was there" or "I felt his presence, definitely, when I was in the booth." The apparitions do not confine themselves to the mirror, or even to the psychomanteum. One man was startled to find the apparitions of three of his departed relatives actually emerge from the mirror and surround him as he sat in the mirror gazing booth. He reached up as if to touch them, wondering at first if we were trying to deceive him by a simple masquerade, whereupon his hand seemed to go right through the apparitions, much to his surprise. A woman felt the presence of her grandfather so palpably that she felt him comfort her in an embrace.

All these encounters, however astounding they might be, seem quite natural while they are going on. They do not seem in any way weird, spooky, or strange. They are certainly not frightening. In every case thus far, it is the relationship between the departed person and the subject that seems to be the centerpiece of the whole encounter. Once the episodes begin, subjects tend to focus on the immediacy of the presence of the departed persons and simply to relate to them and enjoy their presence. Every subject so far has intimated, in one way or another, that for him or her the encounter was experienced primarily as a healing of the relationship with that person."

The "other side":
real help for grief (III)

In this third chapter of my mini-series on "afterlife-based" grief counselling approaches, we look at the fascinating work of clinical psychologist Allan L. Botkin.

Botkin's approach, which he named Induced After-Death Communication (IADC), is an offshoot of the EMDR (eye movement desensitization and reprocessing) technique. EMDR was introduced by Dr Francine Shapiro of California and is referred to by the World Health Organization – together with Cognitive-Behaviour Therapy - as "the only psychotherapies recommended for children, adolescents and adults with Post-Traumatic Stress Disorder." While focusing on the therapist's hand, the patient is asked to move the eyes left or right rhythmically and focus on a disturbing thought, feeling, image, or sensation.

In IADC therapy, people grieving the death of someone or otherwise disturbed by someone's death are asked to focus directly on their sadness during eye movements. Amazingly, the typical IADC session involves the patient reporting having had the experience of a direct, real life encounter with a deceased person, and to have received evidential and comforting messages. In a number of cases, the deceased person relates information previously unknown to the patient.

In a 2005 interview with author Michael Tymn, Botkin said:

> "I discovered IADC by accident in 1995 while working with psychologically traumatized combat veterans at a VA hospital. Our patients, who served in WWII, Korea, Vietnam and Desert Storm,

came to us after reliving the horrors of war in their minds for many years. In 1995 I and my colleagues had been using EMDR for a few years and we had found that we could routinely and rapidly accomplish psychotherapeutic outcomes with EMDR to a degree that we had thought was not possible. In short, we were able to eliminate the reliving component from their memories. I then experimented with a number of variations of EMDR, and I found that a few changes made the standard EMDR technique even more efficient.

Once I began incorporating these changes, I was very surprised when about 15% of my patients reported after death communications during the procedure. Since these patients appeared to resolve their traumatic grief to an even greater degree, I went back through my case notes to find out if I had done something differently when these experiences occurred. I saw that indeed I did do something differently. In the cases in which the experience occurred, I provided an additional set of eye movements without providing any specific instruction to the patient. This additional set of eye movements induced this natural experience. The additional eye movements enhanced what I now call the receptive mode."

The after-death communication experience is unanimously described as very different from all other experiences. In particular, they are obviously different from hallucinations. Technically, hallucinations are perceptions without corresponding sensory input, which means that hallucinations are all in one's head and have nothing to do with any reality that exists separate from us. Hallucinations generally are very negative, vary considerably in content from person to person, and are thought to be a symptom of a severe psychological disorder. It is clear, however, that IADC content is uniformly positive, very consistent in terms of content from person to person, and very healing psychologically. Another observation that indicates that IADCs are related to some reality that exists separately from the experiencer is that there is strong evidence that two people who do the IADC procedure at the same time can have the exact same experience. Botkin called this phenomenon "shared IADCs".

Based on the experience of literally thousands of cases (Botkin later trained his clinical psychology assistants and the approach was then utilized much more broadly, to the point that several dozen trained therapists now operate around the world), other interesting features emerged. First, the IADC procedure works equally well for people experiencing profound sadness or grief and for those with little or no grief at all. Secondly, a strong belief in the afterlife appears to actually have a negative correlation with the likelihood of an experience of after-death communication. Says Botkin:

> "When I first began using IADC I was working with patients in a VA hospital who came to see me to work on their traumatic memories, and were not aware of IADC. When I explained the procedure to them during the course of therapy, a vast majority were very skeptical. Under these conditions, my success rate was 98%. When I retired from the VA and went into private practice, I actually considered offering the guarantee to my clients that they would have an IADC. I figured I could survive financially by not having 2% of my clients pay for their sessions. I was shocked and dismayed when my success rate among people who came to see me for an IADC, and who were not at all skeptical, dropped to about 70%. I am convinced at this point that people who have strong beliefs about the experience have a more difficult time achieving an ADC because their expectations interfere with the receptive mode. ADCs that occur spontaneously come as a surprise to people at times they are not expecting it. (This has been true for me…I wonder why??) A similar psychological state is necessary for an IADC to occur. Of course, the fact that 70% of people who come to have an IADC still have one indicates the power of EMDR to assist one in achieving this naturally receptive state."

However, the most important thing is that IADC provides another example of a "non-traditional" approach to grief counselling which, as I said in the introductory post: a) was developed in a scientific setting; b) is based on or directly implies the existence of life after life; and c) is strikingly effective.

On his website, Botkin summarises nearly twenty years of experience he and his collaborators have had with this method:

> "It doesn't matter what you believe, what we believe, or even what the experiencers believe. The IADC experiences we have induced in thousands of patients result in dramatic life changes that heal grief and trauma in a very short time and are sustained long-term. The technique has worth because it works; it doesn't need for us to agree on a belief system or theory about the source of the phenomenon to support it.
>
> One conclusion is clear: the IADC induction procedure offers the

means to alleviate a great amount of human suffering. There is no greater pain in life than losing a child, a battlefield buddy, or a spouse of many years and feeling disconnected, forever. We lose a part of ourselves when we lose someone so important to us. Now, we can routinely heal this deep pain as well as anger, guilt, and the other emotions resulting from the loss."

The "other side":
real help for grief (IV)

In this fourth chapter of the mini-series of articles about effective but non-traditional approaches to grief recovery, we will explore a method which is as old as mankind itself. A method which is known for working extraordinarily well, but the effects of which are only now being studied scientifically.

The method we are discussing today is very simple: a bereaved person has a sitting with a gifted medium and obtains evidential proof that a deceased loved one goes on existing in a non-material dimension, which we call the spirit world for lack of a better word.

Now, this – the survival of consciousness and human personality to bodily death – is obviously quite an extraordinary claim, and, as famously said by Carl Sagan, it requires extraordinary evidence. This, however, is not the subject of this article. Such extraordinary evidence has been collected for a hundred and fifty years by some of the brightest scientific minds of the planet, including four Nobel Prize winners. When examined with the care it deserves, this evidence leaves little doubt that the extraordinary claim of survival is in fact substantiated. I have produced an eight-hour, adult education video course which reviews in detail this mass of compelling evidence for the specific benefit of the bereaved and the dying, so I will not say any more here.

Rather, what I want to talk about are the effects of an evidential sitting with a medium on the bereaved sitter. This is one of the areas investigated by applied psychical research. Whilst fundamental psychical research is concerned with collecting evidence about extraordinary human experiences and developing theoretical models, applied psychical research investigates the practical applications of the phenomena.

In the case of medium sittings, researching the effects on the sitter may sound a little bit like investigating whether water is wet. After all, obtaining proof of the survival of a loved one as a means to ease the suffering of bereavement is the very reason why people do this. Anybody who has either personally experienced or witnessed after-death communication through a medium knows how extraordinarily effective this can be in reducing grief. Whether in the privacy of a one-to-one sitting, or during a spiritualist church ceremony, or even in one of the showbiz events we see on television, if the medium is good and provides good evidence, people experience major relief.

But, very interestingly, these effects have never been properly studied scientifically. They were in a way taken for granted, resting on the self-evident reality of people's experiences.

It was only in 2010 that this very important subject attracted the attention of serious researchers. Chad Mosher, Julie Beischel and Mark Boccuzzi of the Windbridge Institute carried out an exploratory study showing that – guess what – water is indeed wet!

Using an anonymous survey methodology, 83 participants were asked to retrospectively rate their levels of grief before and after a reading with a medium. As expected, results strongly indicated that participants experience meaningful reductions in levels of grief. But, in my view, the real value of the study lies in the fact that a subset (one third) of participants also worked with a mental health professional (MHP), and were able to draw comparisons between the two approaches.

The participants' verbatim comments about those experiences are extremely telling:

> "The medium had a profound effect on my life and my grieving process... It has helped me in a way I never would have imagined."

> "After the reading, I had a different definition of my relationship with my mom that was more special than I could ever expect."

> "The medium helped me manage the grief that has been with me for more than 20 years".

> "When my first MHP negated the reading I had with a medium, I switched to someone who understood and supported 'my new reality' and therefore received much more constructive help with my grief."

> "I only went to a grief counselor for four sessions. I did not continue because I didn't feel that she was helping me either way."

> "I know that I personally needed to go through counselling as well. However, the level of healing was accelerated by getting readings."

> "The medium reached my heart, the social worker my mind."

The researchers also discussed the advantages of mediumship readings over the frequently reported spontaneous experiences of after-death communication. Their conclusion was that:

> "Readings may be less frightening, less intimidating and easier to understand than more personal, spontaneous experiences. The scheduled and regulated environment of a reading makes it well-suited as a controlled and prescribable treatment option. A medium serves as a non-judgmental participant in the experience who will not disparage or pathologize the experiences of the bereaved".

Whilst the Windbridge Institute is now seeking funding to scale up the exploratory study into a fully-blown research project, we are left once again to ponder over the tragic failures of traditional grief counselling approaches, and on the evident success of non-traditional approaches based on – or implying – the existence of an afterlife.

The "other side": real help for grief (V)

We now come to the conclusion of this mini-series of articles dealing with non-traditional, highly effective and "afterlife-based" or "afterlife-implying" approaches to grief counselling.

In this article I will be talking about my own approach: a method based on the principles of cognitive psychotherapy and essentially consisting in patient education. In a nutshell, I aim to help the bereaved and the dying building a rational belief in the afterlife – a belief based on reason, developed by studying and reflecting on the scientific and empirical evidence, with the ultimate aim to change certain negative, unnecessary and damaging patterns of thought.

This approach is new. I am not aware of anyone else explicitly applying the cognitive psychotherapy technique known as *bibliotherapy* to afterlife knowledge and for the specific benefit of the bereaved and the dying. Therefore, I cannot claim that the approach has been experimentally evaluated. I only have the testimony of the people who have taken the course, who tell me that it is indeed effective in healing the pain of bereavement.

What I will do in this article, then, is to review the theoretical bases and the experience of other researchers which have led me to develop this approach.

I first learned known about cognitive psychotherapy as a patient myself. Some 15 years ago, I suffered a bout of severe clinical depression. For a long time, I lay twice a week on the couch of a Freudian psychoanalyst, crying all my tears in the absence of any – even minimal – progress. Then, in a desperate last-ditch effort, I decided to look around for alternatives. I was living in New York then, and I came across a cognitive-behaviour therapist (an approach less known in Europe at the time) and that literally saved my life - in less than three months I was completely cured of depression.

This experience gave rise to a strong scientific and cultural interest. I read a lot about the theory and practice of cognitive therapy, and, recently, I have also undergone formal training. I was to learn that my own experience is rather typical. Cognitive therapy has been evaluated by some of the largest randomized trials in the history of medicine, and, unlike other forms of talk therapy, has been scientifically proven to be extraordinarily effective.

The foundation of this approach – first developed in the 1950s in the US – is very simple: the way you feel depends entirely on the way you think. By looking at how the brain is wired, cognitive psychologists noticed that a stimulus is first evaluated by thought, and then sent to the emotional centers for the appropriate response. Exactly the same event can give rise to entirely different moods, depending on the way we look at it.

Research has shown showed that depressed people have a "negative bias" – their interpretation of events is pathologically distorted. A mass of negative, distorted, unrealistic thoughts intrude constantly into their consciousness. The solution to this problem is as simple and extremely powerful. Patients do not learn to "think positively" (that has been proven to have no effect). Rather, they learn how to think "non-negatively". Negative, unrealistic, distorted thoughts are recognized and replaced with more balanced, neutral, realistic ones. The effect, I can guarantee you, is simply extraordinary. Honestly, much stronger and immediate than even medication.

Furthermore, this approach does not necessarily need formal sessions with a therapist. Research carried out by one of the developers of cognitive therapy, Dr David Burns, shows that simply reading cognitive-based self-help books and carrying out the exercises with the prescribed modalities can have the same effects of psychotherapy sessions in the case of mild and moderate depression. This particular method is called bibliotherapy, and is essentially what my own grief counselling approach is all about.

Thoughts about the finality of death (one's own, or a loved one's) fuel a large part of suffering for the dying and the bereaved. When these thoughts are shown – through knowledge and reason – to be unjustified, the suffering connected with them can be lifted. This, as you may have heard me saying, will not transform bereavement into a day at the beach. Death and dying come with an unavoidable load of suffering – that is a basic human experience that we simply have to go through. However, correcting the basic thought "death = annihilation" can be powerfully healing.

But that's not all. This is not only a peregrine thought of mine, based on my experience and the knowledge of principles and practice of cognitive therapy. I understood that this approach really had potential when I learned about the research of Dr Kenneth Ring and his collaborators.

As background, it is a well-known fact that all those who have had a near death experience show dramatic and permanent psychological and behavioural changes. They feel a greater sense of purpose, greater self-acceptance, loss of interest for material achievements and, rather, a tremendous thirst for knowledge for its own sake. Most importantly, the NDE literally demolishes the fear of death, completely and forever. While one retains the normal fears associated with the process of dying, the moment of death itself is regarded positively as a liberating transition into a sublime state that NDErs know they have already briefly encountered.

What intrigued me is that Dr Ring and his colleagues showed that some of these psychological and behavioural changes show up in people who have only read about near-death experiences and who have dedicated some time to their study. The more time was invested in learning about these experiences, the greater the changes. And not only that - further analysis revealed that the shifts in values and outlook did not fade with the passage of time. In some cases, these persons were describing changes that had already lasted two decades.

In a nutshell, this, then, is my aim with my course. We know that learning about just NDEs can have such effects. I believe that, by learning that there is evidence from another dozen fields of investigation and that this evidence is completely consistent with the indications from NDE research, the effects can be even stronger.

By correcting the negative, unrealistic (in light of the evidence) thoughts about death, one can indeed transform the fear of death and heal the pain of bereavement.

The powerful illusion (I)

When I talk about my "rational belief" in the existence of an afterlife, and, more in general, when I discuss psychical phenomena, I studiously stay away from trying to provide any explanation. I have no "model", no "mechanisms", no "theory" that could even begin to account for what happens. Based on the evidence, I am convinced that the mind is not dependent on the physical brain and that human personality survives bodily death. Ask me how that happens, though, and I will simply say "I don't know".

Some thinkers and writers do in fact venture to propose theories, mostly pointing to the findings of the so-called "new physics". I must admit, since I have been myself a passionate reader of physics and astronomy before getting the psychical research bug, that these theories generally make my toes curl. I do not claim to be an expert in physics – I am just an "informed amateur". However, most of these thinkers appear to have an extremely sketchy understanding of what quantum physics is really all about (in fact, even the physicists themselves have a problem or two!). What annoys me most is that these proponents take half-digested concepts such as non-locality, quantum entanglement, zero point field and basically say: that's it, here's your explanation for psychical phenomena. A big logical leap, with nothing in between.

Recently I read that an extremely well qualified neuroscientist proposed that psychic phenomena are stronger between people who are related (a well demonstrated fact) because their DNA "resonates" (complete baloney) and that rests on the foundation of superstring theory (a totally fascinating theory, which unfortunately does not have an ounce of experimental support and, anyhow, describes reality at such an infinitesimal level that applying it to huge molecules like DNA is simply bonkers).

Now, where does this leave us, or at least me? As I said, without a trace of an explanation.

However, even if I cannot propose any direct link between the findings of modern physics and psychical phenomena, the world these findings seem to describe is incredibly interesting, mesmerising, and well worth a little "scientific meditation". Also, these findings are a further challenge to the materialist interpretation of reality. In this new mini-series of articles I will be offering a few "pointers" – take them almost as the *koans* of Zen Buddhism. Meditate on them; try to make intuitive sense of them. You will probably not find an explanation, but you may a glimpse that the world is actually very different from what it appears to our senses and common sense; a world in which psychic phenomena seem a little less impossible.

Now, today's subject is: What appears more solid than solid matter? Nothing. And yet, solid matter is not solid at all. Matter, as we all know, is made of atoms, and atoms are basically made of empty space. This is not New Age blabber: quantum physics co-founder Werner Heisenberg, at the beginning of last century, famously said "atoms are not things". Fact of the matter is that an atom is composed of the tiniest speck, called the nucleus, and by a "cloud" of bizarre objects – at times they show up as particles, at times as waves – called electrons, orbiting around the nucleus at such an incredible distance that, if everything was to be scaled up to the size of a football field, the nucleus would be about the size of an ant, sitting in the middle of such field. All around the nucleus, empty space, up to the immaterial "wall" created by the orbiting electrons, which, at this scale, would be where the spectators would sit.

So, atoms are not things. But then, material objects, which are made of atoms - are they things? Well, not really... You are probably thinking I have gone stupid. After all, when you reach out and grab the coffee mug, that feels pretty much like a "thing", very material and very solid. The problem is that you are not touching the mug! The electrons in the outer layer of your skin approach the electrons in the outer layer of the mug, and they, having both a negative electric charge, violently repel each other. You have the sensation of touching an object, but you actually never do it. Believe it or not, a person falling from a height gets killed not by the ground, but by electrical forces. So, materialists believe in things that they cannot touch! Interesting, isn't it?

Furthermore, we are all convinced of seeing objects. The fact of actually seeing the coffee mug is key to our sense of reality, of what is real, material, and what is not. The problem is that the mug in itself is... invisible. We see the mug as its outer layers' atoms absorb photons from the sunlight – or any source of light – and re-emit them. We're not actually seeing the mug, we are seeing sunlight. No sunlight, as far as our senses go, no mug. Here again, materialists believe in things that they cannot see!

Our senses work very well together. They combine the stimuli coming from electrical charges repelling each other producing the sense of touch, with the photons hitting our retina, to give us a perception of the world which is extraordinarily effective for our survival, but essentially false. The only thing that is solid about reality is our perception of it. We will see in the next article how the "real" reality is much better described as a fantastically complex network of relationships, essentially between – and mediated by – energy forms. And, even more strikingly, our unshakeable certainty that the world exists "out there", independently of us, is fundamentally wrong – again, not as a postulate of some New Age basket case, but as scientifically proven beyond any doubt.

The powerful illusion (II)

Amongst the few items that appear with stunning regularity in the accounts of people who have gone through very different kinds of spiritual, transformative experiences is the notion of interconnectedness. Many of those who have had a near-death experience, for instance, report some very similar elements to those who had what they describe as a "mystical experience", either spontaneously or during prayer or deep meditation. These people tell us that they had a direct, immediate realization, based on experience rather than thought or reasoning, of the fact that everything in the universe is linked, connected. In fact, they have felt at one with everything that exists. These are obviously simple words that do not even begin to account for one of the most profound experiences humans can have.

Well, as for myself, I have often read about such experiences, but had never even come close to having one. However, one day I was literally transfixed by an intuitive/scientific thought that gave me at least a shadow of that spiritual realization. Funnily enough, this did not come as I was in some sort of meditation practice, or lost in contemplating some grand natural landscape. I was - hold yourself tight - sitting in a toilet, in the changing room before one of my musical performances (yes, for practically all my adult life I have enjoyed a parallel career as a performing and recording jazz guitarist).

It was there, in this most ordinary situation, as my gaze landed on a patch of sunlight on the toilet's floor, that everything seemed to come together. I simply thought of sunlight, the stream of countless individual photons coming from our star. Having had a keen interest in physics for many years, I knew from my readings that the photons making up the sunlight are produced in the inner core of the sun. The bizarre conditions in there -

ten million degrees in temperature and unimaginable levels of density and pressure - make it so that, even if it travels at the speed of light, it takes 10,000 years for a photon to reach the surface of the sun. From there, the 150 million kilometre journey to earth is covered in a mere eight minutes.

And I thought that the very photons lighting up that room, coming from deep inside the sun, actually were the sun. A physical extension of the very star, reaching out to incredible distances. And, as they hit the room's floor, they got absorbed by its atoms – they quite literally became the floor. And, instantly, they were re-emitted and some of them hit my retina, giving me the visual impression of a patch of sunlight on the floor. Those same photons, produced in the core of the sun, had become the floor, and then they became me! Would there be any better description of this marvellous cosmic dance by which one thing becomes another, and then another, and everything, absolutely everything is interconnected?

And those photons which got absorbed by my retina, and became me, and gave rise to the conscious experience of seeing sunlight... well, those were the sun, becoming conscious of itself!

Proof of hell?

The evidence provided by Near-Death Experiences (NDEs) is one of the pillars of my educational approach to grief and bereavement counseling. The human appeal of the descriptions provided by the experiencers and the quantity and quality of the scientific research around this phenomenon make it easy for me to present it in a convincing manner to the bereaved and the dying as strongly suggestive of the existence of an afterlife.

By looking at the literature and listening to experiencers, two things are immediately apparent about NDEs. On the surface, these experiences present striking similarities. More than 15 common characteristics of an NDE are consistently reported. An NDE may include only one or two of these elements, and, in a few cases, all of them. These include: a sense of being outside one's physical body, sometimes perceiving it from an outside position; a sense of movement through darkness or a tunnel; intense emotions; heightened perceptions; experiencing a great light or darkness; perceiving a spiritual realm, which may include vividly memorable landscapes; encounters with deceased loved ones, spiritual beings and/or religious figures; knowledge of the nature of the universe; a life review; a sense of oneness and interconnectedness; a border of no return; a sense of having knowledge of the future; messages regarding life's purpose. However, when examined in depth, it is clear that no two experiences are identical and no single feature is found in every NDE.

On the one hand, the fact that people with completely different backgrounds have, generally speaking, very similar experiences is one of the strong arguments suggesting that these experiences are real. On the other, skeptics point to differences as an argument suggesting that these experiences are fabricated by the brain.

There are many, very solid arguments, that make the skeptics' explanation untenable, but this is not what I would like to dwell on here. Rather, I would like to reflect a bit more on the puzzle of differences. If, as most researchers directly involved in this field believe, NDEs are indeed suggestive of the continuation of consciousness after death, and therefore of an afterlife, why are the descriptions of the next world partly consistent and partly different? Incidentally, we find the same – and possibly even greater – variability in the description of the afterlife provided by discarnate communicators. The reality the NDErs and the people in Spirit describe appears the same, but it's like it's viewed through very different eyes.

This in itself could be a good explanation. In fact, if you and I went to Paris for a weekend, and were asked to write a one-page description of our visit, many things would be similar, but many things would be completely different. Whilst I see some merit in this line of thinking, I was never completely satisfied. I was recently pushed to do some more reflecting on this by the publication of an excellent book by Nancy Evan Bush (Dancing Past the Dark: Distressing Near-Death Experiences), which shines a much needed flashlight on an aspect of NDEs we have been in some ways "happy to forget about".

The thing is – NDEs are not always heavenly, pleasant, warm, radiant and marvelous. Harrowing experiences are sometimes reported involving similar common elements but with opposite emotional states—extreme fear, isolation, non-being, confusion, occasional torment or guilt. Two substantial studies have reported the percentage of these NDEs as 17% and 18%. The really puzzling thing is that no correlation between the life history, beliefs, behavior or attitudes of a person and the likelihood of having a radiant or harrowing NDE has been established.

Despite this lack of correlation, some, like cardiologist Dr Maurice Rawlings, went as far as saying that negative NDEs are proof of the existence of hell (in the case of Dr Rawlings, going on to say that only conversion to conservative, biblically literal Christianity would save people from that!). Personally, I think that this is baloney. As much baloney, however, as assuming that positive NDEs are proof of heaven.

NDEs, as we understand them today, are not "proof" of anything other than the fact that people report them. However, I am in complete agreement with practically all the foremost NDE researchers who think that they are "strongly indicative" of the survival of consciousness of bodily death.

Where does this leave us, then? Well, to begin with, it would appear that, as discarnate communicators have been telling us consistently for centuries, we do not die. The body dies, but "we" go on. This should come as welcome news for most. And, my recent reflections go, freed from the physicality of the body, our mind/consciousness/personality inhabits a world which is literally created by our own thoughts.

This may be difficult to fathom for some, but it is exactly what has been told to us by countless voices from the Spirit world (I use "Spirit" for lack of a better word to describe the environment in which discarnate personalities appear to exist). In fact, the power of thoughts to create reality is apparent, to a much lesser degree, even in our everyday physical reality (see all the research on psychokinesis or mind-over-matter). But, in non-physical realms, that is pushed to extremes. These are described to us as realms of pure thought – not unlike dreams, only, much more "real" and seemingly "physical". (Appreciate my struggling with words - this is a very difficult a difficult subject to describe!)

So, what is weird about the fact that a Christian will report having seen Jesus during an NDE and a Hindu will tell about one of the many deities of that religion? Or a non-religious person will talk about a "being of light"? What is weird about the fact that people who die suddenly are often said not to have realized that they have died, and remain "stuck" in a reality that resembles closely their life on earth, and in fact physically close to this plane of existence, to the point of making up the large majority of apparitions?

I therefore suspect that, in negative NDEs, something goes awry with the thinking process. Confused by finding itself in an unexpected, non-physical environment, the mind picks on certain memories, thoughts or perhaps hidden beliefs and creates an experience which is sometimes dull or (rarely) frankly harrowing, including images consistent with the Christian hell.

The Buddhist teaching that your state of mind at the moment of death will largely determine what comes afterwards assumes therefore a much more pragmatic meaning. But this is not only a 25 century old wisdom from the East. The need to know that there is an afterlife and prepare for the transition to non-physical levels of experience has been consistently pointed to, over the centuries, by the very people who have already made that transition.

"ATTEND"ing to traumatic grief

I have recently been in touch with a remarkable person, whose research and clinical work has provided me with new and most interesting insights on how we deal with death and bereavement, and made me reflect on how my "afterlife knowledge-based" approach to counselling could seamlessly merge with the mindfulness-based approach that she has developed.

Dr Joanne Cacciatore received her Doctorate from the University of Nebraska-Lincoln and her Masters degree and Bachelor's degree in psychology from Arizona State University. She is the founder of the MISS Foundation and is currently a professor at Arizona State University. Her area of expertise is traumatic grief, specifically following the death of a child. An advocate of "green" mental health care after a traumatic experience, Joanne recently rose to nationwide notoriety in the US on account of her radical – and very well motivated – opposition to changes in the technical guidelines on what is considered traumatic grief and their resulting practical and legal consequences.

Joanne was kind enough to share with me a number of papers in which she outlines her most refreshing approach to traumatic grief counseling, and the preliminary studies showing its beneficial effects. Her approach seems to me the natural evolution of the process that has seen Mindful-Based Interventions (MBIs) gaining, over the last two decades, an increasingly broad range of applications and solid support from clinical research. Whilst mindfulness is a familiar subject to those with an interest in Eastern religions and spirituality at large, as well as to those who engage in non-spiritual meditation practices, it is only recently that it has been the subject of much theoretical and clinical work in psychology. Empirical literature supports the efficacy of MBIs in treating chronic pain, anxiety disorders, irritable bowel syndrome, depression, fibromyalgia, eating disorders and substance abuse. MBIs have also been used to reduce psychological distress among medical students, nurses, caregivers and, critically, survivors of child abuse as well as cancer patients and survivors.

In the first study on the subject, Dr Cacciatore and her colleagues found significant reductions in traumatic stress, anxiety, and depressive symptoms in individuals who had experienced traumatic bereavement and engaged in grief counseling using a mindfulness-based framework. This framework is known as ATTEND, from the initials of its six main components (attunement, trust, therapeutic touch, egalitarianism, nuance, and death education). It is built upon the precept of self-care and compassion, integral for those who work in the emotionally intense fields of social work.

From Dr Cacciatore's own writing, we learn that the model is tripartite in that the client, the relationship, and the clinician experience and benefit from the elements contained in the model. For example, the social worker demonstrates (A)ttunement to the client through attention, awareness, and acceptance of the client's painful emotions. This encourages deepened emotional intimacy with the client, all while the social worker is paying attention to his or her own inner experiences. (T)rust is built through deep listening, empathic caring, and validation of the client. (T)herapeutic touch is used, when appropriate, in a culturally sensitive manner to provide connection, comfort, and presence at times when the client becomes increasingly emotional. An (E)galitarian relationship fosters humble and creative caregiving, addressing each client's unique needs while maintaining role differentiation. In this way, clients guide the intervention based on their own needs and preferences with the gentle companionship of the clinician. (N)uanced counselling pays attention to the individual, familial, and ethnic culture as well as the unique circumstances of each client. In many ways, the social worker becomes a student of the client's culture. Finally, (D)eath education in this model represents the ways in which the social worker becomes educated in death studies and also how psychoeducation can be used as a means through which to empower clients on issues regarding death, dying, and grief when appropriate.

Reading some of the literature on the ATTEND model made me understand that while we take certain things for granted, we absolutely should not. A bereaved person, for instance, would automatically expect a clinician to be attuned and to use an egalitarian, nuanced approach. For a variety of reasons, ranging from the clincian's own stress and "compassion fatigue" to the demands imposed by public agencies' need to work to exacting productivity standards, this is often not the case. With its explicit focus on mindfulness and attention to many essential components in the bereaved-counselor relationship, the ATTEND model certainly has the potential to be of great benefit to both those who are suffering and those who work to alleviate that suffering.

But, naturally, given my stance on the subject, it is the last element, death education, that excited me the most. This is the part of the model in which the compelling evidence for life after life could be integrated - certainly not as an "indoctrination" exercise, aimed at converting the bereaved to the survival hypothesis, but rather as knowledge sharing, with the aim of helping the bereaved reflect on this most important subject, and make an informed decision. As is done in cognitive psychotherapy, disputing the automatic thought "death equals annihilation/disappearance" on the basis of empirical evidence and rational evaluation can be of extraordinary benefit to somebody who's lost a loved one. "Afterlife education" could also help the many who have anomalous experiences surrounding the death of a loved one and don't know what to do with them, how to integrate them into an imposed worldview dominated by materialism. The sheer fact, for example, of getting to know how frequent and widespread these experiences are can help a person feel less isolated, less "mad" or "visionary", better accept his or her own experience and use it as part of the recovery process.

Do mediums really talk to the dead? (I)

For many bereaved persons, the only consolation is the belief that a deceased loved one goes on living in a non-material dimension of existence. Demonstrating that such a belief is not a matter of blind faith, or wishful thinking, but can be entertained by a rational person, based on masses of compelling evidence, is my own approach to grief counselling. Showing that the automatic thought "death = annihilation/utter disappearance" is in fact not true does not fill the void left by the loved one's passing, but removes at least one significant layer of suffering and hopelessness, with remarkable positive results for the bereaved.

However, it is one thing to be rationally convinced of the survival of consciousness and personality of bodily death, based on the study of the available evidence. Quite another thing is knowing, based on direct, personal experience. For thousands of people all around the world, such direct knowledge comes through the services of a gifted medium. From the perspective of the sitter, the bereaved person, an evidential sitting with a medium amounts to living proof. He or she will not need anything else. The rest of us will have to rely, once again, on reason. A big question, then, begs to be asked: Do mediums really talk to the dead?

The short answer is yes.

The long answer, if you want to satisfy your reason and rationality, is a lot longer. In this mini-series of articles, I will summarize the methodology and results of the multiple lines of research that have brought, during the last two decades, mediumship research into the laboratory.

Confronted with a claim as preposterous as communicating with disincarnate personalities, our rational mind would come up with at least three key questions:

First - are the statements by the mediums precise, focused and specific enough to actually mean something for the "sitter" (the person who consults the medium to contact a deceased loved one)? This is the criticism typically levelled by skeptics, who claim that statements by the mediums are intentionally vague, so that anybody - especially a bereaved person - can read something into them.

Second - if the statements are indeed specific and meaningful, can mediums deliver them without any previous knowledge about the disincarnates or sitters, in absence of any sensory feedback, and without using fraud or deception? There is no need to say that these are the "weapons" used by the skeptics to try to discredit what we technically define as Anomalous Information Reception (AIR).

Finally - if AIR actually takes place (which in itself I find most extraordinary and fascinating), does this information actually come from disincarnate personalities? Observers of this phenomenon have proposed that such information may be "read" by the medium from the mind of the sitter, or from some sort of "memory" embedded in the fabric of space-time.

To answer these key questions, psychical researchers have employed the most sophisticated investigation techniques since the end of the 19th century. Through exquisitely refined research protocols and with the most rigorous controls, "historical" researchers have undoubtedly answered "yes" to all the three questions. However, this was "field" research - observing and trying to understand a phenomenon as it happens in its natural environment. It was not until some 20 years ago that mediumship research has been brought into the laboratory, an environment in which investigators have control of most - if not all - the variables of the process, and can perform quantitative, statistical analysis on the results, as it happens in any other branch of science.

In this mini-series, we will discuss several experiments carried out by different groups on both sides of the Atlantic, looking at how protocols became increasingly sophisticated and tight in order to eliminate any possible weakness. And, we will see how the increasingly stringent controls did not affect the results. The answer to the three questions is still a resounding "yes". As far as I am concerned, Anomalous Information

Reception is proven. Period. I will also briefly discuss alternative explanations, showing that the fact that mediums actually do talk to the dead is the least unbelievalble explanation for the evidence.

Before I conclude this brief introduction, one word of caution, and one pointer: The fact that mediumship has been proven in the laboratory does not mean that it is a common faculty amongst human beings. To the contrary, research shows that most of us do not, under normal circumstances, have this capacity. A few individuals have it to some degree, and precious few are "superstars", with an extraordinarily more developed capacity to contact the spirit world. Whilst we will see how research has dealt with this uneven distribution of talent, beware of those who, in the real world, may take advantage of people who find themselves in difficult situations, such as the bereaved. As a rule of thumb, real mediumship is not a commercial affair. Real mediums do not charge money for their sittings, or they charge minimum amounts. If you want to go and see one, make sure you do your research.

The pointer is just one example that, for me, makes all this science and this philosophizing pointless. Just look at the extraordinary Scottish medium Gordon Smith talking to a couple of bereaved parents, as filmed in a BBC documentary. What does this tell you, as a human being? Imagine that the video has been posted by somebody as an example of "cold reading tricks". How sad is that?

The video is available at:

> https://www.youtube.com/watch?v=WDJhVkFxiKA

Do mediums really talk to the dead? (II)

The knowledge, provided by a sitting with a gifted medium, that a deceased loved one has not simply "vanished" out of existence, but he or she remains in fact very present - although not normally visible - in the lives of those left behind, is of extraordinary comfort to the bereaved. As part of my educational approach to grief counseling, I discuss a range of scientific experiments that indicate that medium sittings are indeed what they appear to be - communication with discarnate personalities.

In this second article of the mini-series we will look at the investigations of Prof Archie Roy of Glasgow University and fellow physicist Tricia Robertson. The original scientific papers were published in the Journal of the Society for Psychical research in January 2001, April 2001 and January 2002.

Firstly, the skeptical hypothesis that statements made by mediums to recipients are so general that they could as readily be accepted by non-recipients was tested. A two-year study involving 10 mediums, 44 recipients and 407 non-recipients ostensibly falsified that hypothesis. Although the original paper discusses a number of methodological intricacies to eliminate fraud and errors, the basics are simple. A medium has a public sitting with a number of people, called *recipients*. Only the statements made by the medium to individual recipients are written down. The statements are shown to the recipients who either accept them or refuse them. The same statements are subsequently shown to a control group who were not present at the sitting and who are called *non-recipients*. The members of the control group either accept them or refuse them. No matter where the experiments were carried out, no matter the manner of their recording, no matter whether the information given by the medium was at a public meeting or a smaller group, the results were consistent in that there was a large gap between the set of statements accepted by the recipient and by the non-recipient.

Secondly, the researchers published the details of a strict protocol which would eliminate the possibility for the medium to get indications from body language and verbal responses from the recipients. A randomized seat numbering system was also implemented, which means that the experimenter who actually numbers the seat cannot possibly know who will sit in any particular seat. Along with this, the experimenter who analyzes the initial data does not know which seat numbers have been pre-selected, therefore there is no knowledge of which recipient sat in which seats.

Thirdly, the authors go on to apply this new strict protocol to a carefully designed set of experiments. These are also designed to isolate factors, such as "Will a person accept more statements if they think or know that they are actually the recipient?" "Will a person accept fewer statements as relevant in their life if they think or know that they are not the intended recipient?"

This third paper covers 13 different experimental sessions carried out throughout the UK, with participants always gathered by a third party. The average number of participants at a session was approximately 25. Typically six experiments were carried out at each session. The authors identified 15 categories of participant through a 2 character code. The 1st character is an 'R' (recipient) or an 'N' (non-recipient). The 2nd character is lower case and indicates what the participant believes their role to be. For example, a recipient who believes that he/she is the recipient and who is *actually* the recipient would be designated by the symbols Rr. An actual recipient who believes that they are not the recipient is Rn. A recipient who does not know whether or not they are a recipient would be Rq. There is also a category P, which is used in the experimental sessions where no actual medium has been used (although the audience thinks that there is a medium). This allows responses to be analyzed with no possibility of psychic interference from a medium.

Using statistical analysis the authors were able to evaluate the responses of every category and examine the effects, if any, of psychological factors. Even under triple (arguably quadruple) blind conditions, the intended recipients' acceptance levels continued to be higher than non-recipients, the odds against chance being a million to one.

Very interestingly, the results incorporate all of the mediums who participated in the research; if the authors had only given the results from the "superstars", the odds against chance would have been even greater.

Do mediums really talk to the dead? (III)

Gifted mediums provide the bereaved with evidential proof that their deceased loved ones have not simply vanished out of existence. Rather, although not normally perceptible by the senses of the living, they appear to be very much present in our lives, attentive to and conscious of whatever goes on around us. This is what countless sittings seem to indicate, and the bereaved themselves are in no doubt about their therapeutic value. However... Is this just a fantasy, a fairy tale people are happy to accept as reality to ease their grief? Are the bereaved reading too much into what are simply general statements made by mediums? Are mediums "cold reading" the sitters to extract non-verbal information? In this mini-series of articles we are looking at what science had to say, during the last two decades, in response to such questions.

No discussion of laboratory-based mediumship research would be worth much without mentioning the work of Prof Gary Schwartz, Department of Psychology, University of Arizona. Over the years, Prof Schwartz and his collaborators (particularly Julie Beischel, who later moved on to co-found the Windbridge Institute) have consistently reported highly significant results indicating that "anomalous information reception" indeed takes place, using increasingly sophisticated experimental protocols which effectively rule out the possibility of fraud, deception, sensory leakages and even telepathy as possible explanations.

It is important to stress that, similar to the studies by Prof Archie Roy and Tricia Roberston in Glasgow, whilst employing the most stringent protocols to eliminate conventional explanations, these experiments were also designed to maximize the potential for positive results. In particular:

a) research mediums were selected after having demonstrated that they were able to perform accurately under normal mediumship or single blind conditions; b) sitters were selected to be highly motivated to receive information purportedly from their deceased loved ones and thus score the readings accurately; c) scoring systems fostered detailed item-by-item analysis of the readings, followed by meaningful summary scoring; and d) experimental conditions optimized the mediums' potential to receive information. In the words of Julie Beischel: "If you want to see whether a seed produces a plant, you recreate the natural conditions in which this process occurs – you do not put the seed in cement in sub-freezing temperatures…"

I will now briefly describe one of the last sets of experiments (*Anomalous Information Reception by Research Mediums Demonstrated Using a Novel Triple Blind Protocol*) published in 2007 by the scientific magazine Explore. In this study, eight University of Arizona students served as sitters: four had experienced the death of a parent; four, a peer. To optimize potential identifiable differences between readings, each deceased parent was paired with a same gender deceased peer (for instance, a deceased grandmother with a deceased female peer). Eight mediums who had previously demonstrated an ability to report accurate information in a laboratory setting performed the readings. Sitters were not present at the readings; an experimenter blind to information about the sitters and deceased served as a proxy sitter. With this backdrop, the procedure went as follows:

1) PAIRING. Information about each discarnate and his/her relationship with the associated sitter was collected from the sitter participants by a research assistant who did not interact with the mediums. Discarnate descriptions were then paired to optimize differences in age, physical description, personality description, cause of death, and hobbies/activities of the discarnate. Four deceased parents were paired with four deceased peers of the same gender for a total of four pairs of sitters (for example, a grandmother who had died of heart failure, was short and loved gardening was paired with a female student who died in a car accident, was tall and played the cello).

2) MEDIUM READING. Each of the eight mediums performed two readings: one for each sitter in a pair. Each of the four pairs of sitters was read by two different mediums for a total of eight pairs of readings. The mediums were given no information about the sitter or his/her relationship to the discarnate. However, to increase the capacity of the medium to receive accurate information about a targeted discarnate, the first name of the discarnate was given to the medium at the start of the

reading.

To optimize testing conditions, the mediums performed the readings over the phone at scheduled times in their homes. The digitally audio-recorded phone readings took place long distance; the medium was in a different city (if not state) than both the absent sitter and the experimenter acting as the proxy sitter.

Each reading included three parts: a) Deceased-Directed, in which the experimenter gave the medium the first name of the discarnate and asked the medium to receive and report any information from the discarnate; b) a Life Questions procedure, in which the medium was asked four specific questions about the named discarnate's physical appearance, personality, hobbies, and cause of death; and c) a Reverse Question condition, in which the experimenter asked, "Does the discarnate have any comments, questions, or requests for the sitter?"

3) ITEMIZATION. Each reading was transcribed and a corresponding numbered list of individual items (i.e., separate, stand-alone pieces of information) was created by an experimenter blind to details about the sitters or discarnates.

4) SCORING. Each sitter scored the reading intended for him/her as well as the reading of the control sitter while remaining blinded to the origin of the readings. The readings were scored according to the following scale:

6: Excellent reading, including strong aspects of communication, and with essentially no incorrect information.

5: Good reading with relatively little incorrect information.

4: Good reading with some incorrect information.

3: Mixture of correct and incorrect information, but enough correct information to indicate that communication with the deceased occurred.

2: Some correct information, but not enough to suggest beyond chance that communication occurred.

1: Little correct information or communication.

0: No correct information or communication.

5) CHOOSING. After summary scoring was completed for both readings in a pair, the sitters were asked to "Pick the reading which seems to be more applicable to you. Even if they both seem equally applicable or non-applicable, pick one." They were then asked to rate their choice compared to the other reading according to the following scale:

a. clearly more applicable to me

b. moderately more applicable to me

c. only slightly more applicable to me

d. both seemed applicable to me and to the same extent

e. neither seemed applicable to me

Now, let's look at the results. On the whole, the average summary rating (point 4 of the procedure above) was significantly higher for the intended readings than for the matched controls. It is noteworthy that three mediums produced dramatic findings with summary scores of 5.0 and 5.5; two mediums produced moderate findings (summary scores of 3.5); and none of the mediums produced reversals (i.e. control ratings higher than intended ratings).

When asked to choose which reading was more applicable to them, sitters chose the readings intended for them 81% of the time. Of those 13, seven were rated "clearly more applicable" and three as "moderately more applicable".

Do mediums really talk to the dead? (IV)

Now, we complete this mini-series on laboratory mediumship research with another interesting study carried out by Emily Williams Kelly and Dianne Arcangel, and published in 2011 by the prestigious Journal of Nervous and Mental Diseases.

The substance of the study is pretty much the same as in the experiments we have described earlier. Nine mediums and 40 sitters took part in this exercise. The two authors acted as proxy sitters, so that the mediums were blind as to who the intended recipient was. The mediums were given just a photograph of the deceased person, and nothing else – no name, no age at the time of death. The mediums' readings were taped and then transcribed, removing any references to the appearance of the person in the photograph or any other such clues. Each sitter was then sent 6 transcripts—the real one, as well as 5 intended for other persons, all 6 selected from the same age and gender group. The sitters ranked the transcripts from 1 to 6, 1 being the most accurate.

As in previous studies, the results were highly significant: 14 of the 38 readings were correctly chosen, and 7 others were ranked second. Altogether, 30 of the 38 were ranked in the top half. Statistically, the probability of obtaining such results by chance is less than one in 10,000. This study indicates again that mediums are indeed capable of providing information about a deceased person whilst all known channels of communication are closed, and that the results were not due to chance or interpretation of generic statements.

However, what is really interesting in this particular study is that the authors have looked not only at the quantitative aspects, but also at the qualitative ones. These are equally important, not only because of the emotional impact they often have on the bereaved person, but also because they may give us more insight into the kind of information more likely to come through.

Most of the 14 people who correctly chose their own reading made comments such as "I don't see how it could be anything other than (X reading)"; "I feel certain this is the correct choice and would bet my life on it"; "one reading stood out from the rest ... I just know [it] was correct because it sounded like my mom"; "it had the most instances that could apply to my son."

In addition to such general statements, however, some did go on to comment on specific details that impressed them. For example, the person who "would bet my life" on his choice cited the medium's statement that "there's something funny about black liquorice . . . .. Like there's a big joke about it, like, ooh, you like that?" According to the sitter, his deceased son and his wife had joked about liquorice frequently. Also, the medium had said "I also have sharp pain in the rear back of the left side of my head in the back, in the occipital [bone]. So perhaps there was an injury back there, or he hit something or something hit him." The deceased person had died of such an injury incurred in a car crash.

In another reading, the medium said "I feel like the hair I see here in the photo is gone, so I have to go with cancer or something that would take the hair away," and later "her hair — at some point she's kind of teasing it, she tried many colours. I think she experimented with colour a lot before her passing." The girl's mother confirmed that she had died of cancer, had dyed her hair "hot pink" before her surgery, and had later shaved her head when her hair began falling out (her hair was normal-looking in the photo).

In another example, among many other details the sitter commented especially on the statement "he said I don't know why they keep that clock if they are not going to make it work. So somebody connected directly to him has a clock that either is not wound up, or they let it run down, or it's standing there just quiet. And he said what's the point in having a clock that isn't running? So, somebody should know about that and it should give them quite some laughter." The sitter did laugh (and cry) over this, because a grandfather clock that her husband had kept wound had not been wound since his death. The medium had also commented that "he can be on a soap box, hammering it". His children when young had frequently complained about "Dad being on his soap box".

The article cites many other such examples. These kinds of anecdotes are nothing new to anybody who has studied mediumship or had a sitting with a good medium. I very much welcome the fact that these particular ones were published in a top peer-reviewed scientific publication, along with convincing statistical results.

And, yes – it does indeed appear that some gifted mediums really talk to the dead.

Afterlife science in support of the bereaved

A few minutes after midnight on January 1st, 2014, I received a very welcome email message, announcing that I had been appointed to the scientific advisory board of the Forever Family Foundation in the U.S. This really added an edge to my New Year celebrations. In order to appreciate the extent of my elation, you must understand a couple of things.

First of all, I still consider myself a "freshman" of psychical research and afterlife studies. I have not been in this field for even ten years, yet I find myself joining the company of some of the very intellectual giants who have inspired me. Seeing my name on the same list of people such as Bruce Greyson, Gary Schwartz, Dianne Archangel, Julie Beischel or Stephen Braude is an enthralling experience, even for a "low-ego-quotient" person like me.

Secondly, and this is way more important, I am very excited to be formally associated with an entity which embodies, at the level of an entire organization, my very own approach to working with the bereaved and the dying.

For those of my readers who yet don't know it, the Forever Family Foundation is a science-based, all-volunteer, not-for-profit, non-sectarian organization that has just reached its 10th year. This charitable foundation:

- Supports the premise that life does not end with physical death
- Furthers the understanding of afterlife science and survival of consciousness

- Offers support to the bereaved.

Among the members or the organization and its various Boards are scientists, researchers, medical doctors, philosophers and educators who have devoted substantial parts of their careers to the investigation of the survival hypothesis – an existence beyond the physical world. The basic premise of the work of the organization – exactly as it is for my own work, on a much more limited scale – is that a rational belief in the afterlife (based not on blind faith but rather on knowledge and critical review of the masses of compelling empirical evidence in support of the survival hypothesis) can be of extraordinary benefit to those who have lost a loved one as well as to those who are facing death.

Membership in the global organization, as well as its core services, is free of charge. The Foundation maintains an informative website, publishes and mails comprehensive newsletters to members, hosts a weekly radio show entitled "Signs of Life", conducts Afterlife Discussion Groups, and directs a Medium Certification Program. In addition to its free services the organization holds seminars, lectures, workshops, demonstrations, grief retreats and conferences.

I must admit, however, that I found the personal contact with the two founders of this initiative at least as magnetically attractive as the mandate and scope of work of the entire organization. I will not mention their names here, as, in our dealings so far, I sensed a refreshing, endearing taste for understatement and reservedness on their part. They were kind enough to share with me brushstrokes about their personal background and about the circumstances that led to the creation of the Foundation, and all this greatly contributed to my enthusiasm in being associated with this undertaking.

On my part, I look forward to making whatever little contribution I can to the work of the Foundation. And, with all my heart, I encourage my readers to join this 10,000 member-strong worldwide community, support its mission and spread the word as widely as possible.

Guess what – paranormal raps are... paranormal!

A considerable part of my learning about psychic phenomena comes from the pages of the Journal of the Society for Psychical Research (SPR), a scholarly publication dating back to 1882, which publishes, after a rigorous peer-review process, papers from high-caliber investigators. The May 2010 (Vol. 73) edition of the Journal carries an article by scientist Dr Barrie Colvin, Ph.D. which I found particularly interesting and want to bring to the attention of a wider public. Before I do that, though, I would like to warmly encourage my readers to consider joining the SPR. Just the publications (the Journal, the Proceedings and the Paranormal Review) would be well worth the modest yearly membership fee. There are other, tangible and intangible advantages in becoming a member, however, and anybody with a scientific mindset and an interest in the paranormal should consider joining. More information at www.spr.ac.uk.

Now, the article I would like to talk about in this post shows instrumental evidence for an inexplicable and objective banging sound detected in recordings made during alleged poltergeist activity.

Whilst paranormal rappings associated with poltergeist activity have been described for many hundreds of years, it is only recently that an interesting pattern has been detected within the fine detail of the paranormal rapping sounds. Whereas raps and knocking sounds produced by ordinary means exhibit a normal acoustic pattern, those recorded in alleged poltergeist cases show quite a different sound signature. No explanation can be found for this pattern at present.

Dr Colvin has analyzed recordings of alleged poltergeist knocking obtained from around the world over a 40-year period. The earliest was a recording made by a local physician at Sauchie (Scotland) in 1960 and the most recent was obtained from a poltergeist case at Euston Square, London in 2000. The sample involved 10 separate recordings recorded on different recording apparatus.

Whilst the two types of raps sound rather similar, computerized analysis shows that they are actually acoustically different. In each of the allegedly paranormal recordings, regardless of the place or time of the recording, and even of the recording apparatus, a particular sound pattern is detected which so far remains unexplained. Attempts to replicate this pattern in ordinary ways have so far been unsuccessful.

The essential difference between these raps and those produced by normal means lies in the details of their sound envelope (the way sound evolves over time). In the case of a normal rap, the sound (which often only lasts a few milliseconds) starts loudly and decays over a period of time. The loudest part of the sound is right at the beginning. In the case of a poltergeist rap, the loudest part is *near* the beginning of the sound - but not at the very beginning. The rapping sound starts relatively quietly and works up to a maximum before it then starts to decay. This effect has been seen in all ten of the poltergeist cases studied.

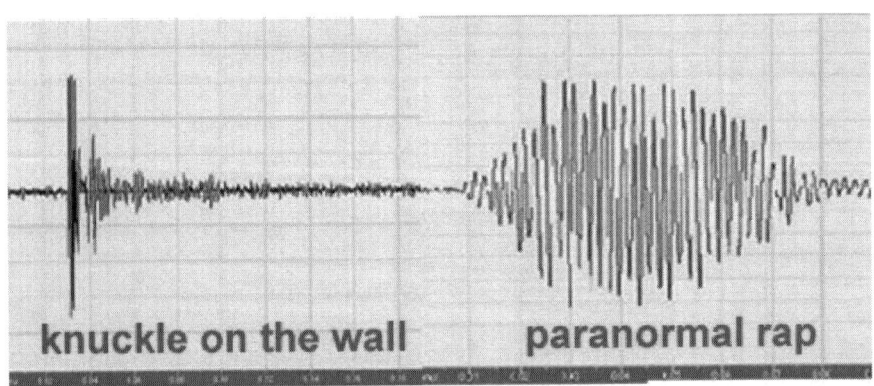

The picture above shows the waveform of a sound produced normally, on the left, and on the right the waveform of a paranormal rap.

The question arises as to how such a sound is generated. There is evidence which points to the sound arising from within the structure of a material rather than from the surface of it, as would be the case with a normally-produced rapping sound. This phenomenon will be examined further in future investigations of poltergeist activity.

Dr Colvin states: "Ever since my personal involvement in the investigation of a rapping poltergeist at Andover, Hampshire, in which it was absolutely clear that no normal explanation could account for the observed phenomena, I wondered whether the recorded raps were in any way different to those produced by normal methods. It is now clear that they are indeed different".

Mind is not in the brain

In order to begin to open up to the idea that what we call "ourselves" – our mind, our consciousness, our personality – goes on living after the death of the physical body, we first have to understand that this "ourselves" is related to but – crucially – independent from the physical brain.

The relationship between mind and brain is one of the most hotly debated subjects in the fields of philosophy and natural science. So hotly, in fact, that it provides the ground for nothing less than a true war of religion. On one side of the frontline, the fashionable and currently dominant theory of materialism maintains that all that exists is matter. Mind simply does not exist – it is an illusion, produced by the electrical activity of the brain. You, as Richard Dawkins famously said, are just a biological robot.

On the other side of the frontline, we find scientists and thinkers who believe that science is not a set of fixed ideas but a method, and the method says just one thing: follow the data, follow the evidence, and base your understanding of the world on that. They look at the experimental data and conclude what I said in the beginning – mind is related to, but independent from, the activity of the brain.

Materialists also follow an extreme interpretation of the so-called scientific method – if something is not published in peer-reviewed scientific journals, it either does not exist or is not true. This is interesting, in a way, because Tatties, the cat that sits on my desk as I write, has never appeared in a peer-reviewed journal, and still I am pretty sure she exists…

You may then conclude that materialists do not consider evidence for an independently existing mind because it does not appear in peer-reviewed journals. How wrong! There are literally thousands of scientific papers, published in all sorts of peer-reviewed journals, which indicate that materialism is a false theory. Mind is related to, but not the "product of", the electrical activity in the brain.

Let's take just one example, but not one chosen at random. I pick this particular one because it concerns one of the most cherished technologies used in support of the mind = brain theory – fMRI.

Functional magnetic resonance imaging, or fMRI, is a technique for measuring brain activity. It works by detecting the changes in blood oxygenation and flow that occur in response to neural activity – when a brain area is more active it consumes more oxygen and to meet this increased demand blood flow increases to the active area. fMRI can be used to produce activation maps showing which parts of the brain are involved in a particular mental process.

Materialists point to an ever increasing mass of studies showing the links between all sorts of mental processes and highly specific fMRI findings and say – Hey, what more proof do you want? Mind is in the brain.

This is also interesting, because it is the same as saying: I have measured the voltage of a transistor in my FM radio, and it changes when I hear a voice through the speaker. So, hey, what more proof do you need? The transmission originates in my FM radio!

Never mind, let's not get distracted. Let's look at the results of a most interesting 2005 study (*Evidence for Correlations Between Distant Intentionality and Brain Function in Recipients: a Functional Magnetic Resonance Imaging Analysis*) published in the Journal of Alternative and Complementary Medicine – a *very* peer-reviewed scientific journal. The paper's abstract says:

> "This study, using functional magnetic resonance imaging (fMRI) technology, demonstrated that distant intentionality (DI), defined as sending thoughts at a distance, is correlated with an activation of certain brain functions in the recipients. Eleven healers who espoused some form for connecting or healing at a distance were recruited from the island of Hawaii. Each healer selected a person with whom they felt a special connection as a recipient for DI. The recipient was placed in the MRI scanner and isolated from all forms

of sensory contact from the healer. The healers sent forms of DI that related to their own healing practices at random 2-minute intervals that were unknown to the recipient. Significant differences between experimental (send) and control (no send) procedures were found (p = 0.000127). Areas activated during the experimental procedures included the anterior and middle cingulate area, precuneus, and frontal area. It was concluded that instructions to a healer to make an intentional connection with a sensory isolated person can be correlated to changes in brain function of that individual."

Now – what's going on here? Simple: a scientific study using the materialists' most revered technology – and published in a peer-reviewed journal – proves that they are wrong. Mental activity in one person determines measurable changes in the brain of another person when all communication channels are closed. This a special form of psychokinesis or "mind over matter", well known and very well documented for many years. It demonstrates with no ambiguity that mind and matter are definitely not the same thing.

What do materialists say? Deafening silence. They simply ignore the evidence. If they cannot ignore it, they try to suppress it (see, for instance, the ugly Wikipedia wars…). And, as a last resort, they insult, ridicule and smear researchers and proponents (see, for instance, the entry about myself on RationalWiki).

What do the rest of us do? We understand that mind in not dependent on the functioning of the brain, and we open up to the fact that, in a way which we do not understand, "we" survive the death of our physical body.

Assisted After-Death Communication (I)

In 2012, I had the extraordinary experience of training personally with one of my intellectual heroes, Dr Raymond Moody. Raymond shared with me his experience and insights about the technique of "assisted after-death communication" he developed in the 1990's, which he called psychomanteum. This ancient Greek name refers to a darkened-out room in which a person seats comfortably on a recliner and gazes calmly into a mirror reflecting what I could only describe as "faintly luminous nothingness". As shown by Dr Moody and other researchers, most of the people who sit in the psychomanteum following an in-depth psychological preparation do have an experience of reunion with a departed loved one. Such experience can vary from a generic but unmistakable sense of presence to full visual apparitions with sustained two-way communication.

Being keenly interested in the therapeutic applications of this procedure, I had thought of experimenting with it myself, but then I was discouraged by the difficulties of actually building a room with the necessary characteristics. I then turned to an alternative method – the one initially developed by Allan Botkin, a clinical psychologist who reported extraordinary and consistent results applying a technique called Eye Movement Desensitisation and Reprocessing (EMDR).

For a few months I worked with volunteers, combining the preparatory psychological techniques I learned from Dr Moody with the application of eye movements as described by Allan Botkin and others.

Although the numbers were low, the results, I am glad to report, ranged from good to frankly stunning.

Some of my subjects had a full experience of reunion, and that in itself I find extraordinary. But, as a medical doctor and psychotherapist, I am mostly interested in the therapeutic potential of this method, and the results in this respect went beyond my expectations.

Let's look at what my first subject – female, mid-fifties – wrote about her experience.

"**The images**

There is a heart-shaped trellis covered in pink/white roses. It is the entrance to a vast sunlit landscape of green into infinity.

The trellis entrance remains in daylight while the sky of the landscape turns a midnight blue and is densely star studded.

I see my mother's face – she has a beaming smile and I note on her face the kindness and gentleness I remember so well. She is wearing glasses, and to my surprise and happiness, she has her "pre-Alzheimer eyes". This is the first time I see these eyes since my mother's passing (when she has appeared in all my dreams since her passing it has always been as someone with Alzheimer's). She is showing me opaque white steps with a lilac misty tinge - no railings on either side - leading to a higher place. Her smile is genuine alongside these steps.

The image at the very end is powerful. There is a swooping/fluid figure gathering up sand.

The woman is younger and has a head covering – she does not look like my mother but I know and understand it is a younger version of my mother, even although I do not recognize her. Rising up dramatically (with a whooshing sound), she spreads her arms and all the sand is falling through her outstretched fingers – at the same time the figure is disintegrating the sand settles like dust to the earth, into nothingness. It as if the episode is now "done and dusted". The related feelings of my own guilt, regret, sadness and helplessness seem like they have been looked at and processed and, although not gone completely, greatly diminished.

What I got out of it:

- Feeling of peace.

- Letting go – churning in the stomach when recalling the events practically gone at the end of a 2-hour session. When I think of it today, there is a pang but not the same sickening sensation.

- Permission to let the guilt go.

- Forgiveness from another place.

- Self-forgiveness.

I woke up this morning with a lighter heart – not the usual feeling of heaviness and struggle as the day begins."

Now, as a clinician I can only be stunned. Self-reported improvements like these would be considered a full success after a full course of psychotherapy – a minimum of 12 weeks with the shortest protocols. In this case, they were obtained in a single, three hour session!

Assisted After-Death Communication (II)

In this article I will attempt to describe a truly extraordinary session – certainly the most striking in the small series of experiments I carried out so far. The subject, a man in his early fifties, volunteered as a test subject mostly out of curiosity. We had agreed that I would tell him nothing about what he may experience, and that he would not do any reading/finding out on his own. During the in-depth talk we had as part of the preparation immediately prior to the session, we did unveil a couple of emotional "trouble areas", as well as a few mildly traumatic memories, linked to the death of his parents, who had both died within a year of each other, about a decade earlier.

Having found good psychological material to work on, we started the eye movement part of the experiment by focussing on one of the trouble areas and one particular traumatic memory. For about an hour we alternated between eye movements and discussion, loosely mirroring a standard EMDR session. The subject gained some interesting and reportedly useful insights about his emotional problems, but no experience of reunion happened.

When I felt that we had made some progress on one issue, I gave the subject the choice of concluding the experience or continuing and tackling the other main problem/memory. He wanted to continue, so we did.

I asked him to focus on the second traumatic image/memory and I conducted the usual 24 eye movements. Then I asked him to let that go, close his eyes, and tell me if anything came up. After a while, he said "This is very strange. I see that table where we used to eat in the kitchen when I was at home with my family." Sensing that perhaps we were on to something, while administering more eye movements, I asked him to focus on that image. I asked him "Think of that table – its height, the color of the tablecloth, where the light would come in from, where the other members of the family would be sitting". Then, again, I asked him to let that go, close his eyes, and tell me if anything came up.

He lay back on the armchair for perhaps five or six minutes, eyes closed, without saying a thing. Then it was like he had been stung by a bee, or scared by something. He opened his eyes, almost jerked up from the recliner, and for a few moments he looked around as if completely lost. Then he relaxed and broke down in tears. This is what he wrote the next day:

> "When I closed my eyes, focusing, I saw an image of the dining room table in the house of my childhood. The perspective was of a side view of the corner of the table and I could see clear details of the wood grain and even smell the polish. I focused intently on the table and then I could see the black silhouettes of a number of relatives who used to come around to our house during the Thanksgiving and Christmas holidays. Slowly, in the upper left of my field of vision, two figures appeared. These looked like oblong water color blotches, orange-brown in the center diffusing to light yellow around the edges. Masses of energy is how I would most accurately describe what I was seeing, but for the fact that these were very clearly my mother and father, with my mother's figure slightly closer to me than my father's. Although I could not hear their actual voices, both were speaking to me at the same time. They said "we love and support you, son, and always will. You are a good father and you have two beautiful children ". As I write these words down it strikes me that what they said was very short, but at the time, and even now, these words feel to me to be fully complete in terms of the feelings they convey. In any case, at this point in the exercise I was literally overcome with emotion and broke out in what I would describe as tears of love and joy. I'm not sure how long the entire episode lasted, perhaps a few minutes at the most, but the impact on me was that of a deep sense of being loved. I also have an abiding sense now that my parents "know" my children, which is a wonderful feeling. I have no doubt that this experience will be one from which I will draw strength for a long time to come."

It took this grown-up man – a senior international civil servant with the United Nations – quite some time to recover his composure. After a while, he stood up from the armchair and hugged me tight, literally sobbing. At that stage, I must confess, I was in tears myself. I was swept away by the happiness of being a part of this extraordinarily meaningful event. Meaningful for him, certainly, but for me as well - it had happened, exactly as I had read in the books, and I was there to witness this, perhaps I had helped bringing this about...

What can be said about such an experience? As I see it, there are two possible explanations.

Number one: after their death, the father and mother of the subject went on living in a non-physical dimension of existence. Through this procedure, they managed to convey a strong sense of presence and a very meaningful message to their son. This explanation is consistent with the colossal amount of empirical evidence which supports my rational belief in an afterlife.

Number two: somewhere hidden in our minds exists a "doctor", a powerful healer who is also a magician – something which has the capacity of conjuring up in no time a complex hallucinatory experience which conveys: a) a full sense of reality; and b) a highly specific, targeted and powerfully healing message (the emotional problem the subject and I were focusing on had nothing to do with what was said, but the subject told me "these were the exact words that I needed to hear").

I find both explanations utterly fascinating.

Mind and Molecules

What has an alpinist atop Mt Everest (this was shown in a picture on the website where the original article appeared) to do with the suffering caused by bereavement? A lot, as we will see, but this requires a bit of explanation.

My approach to grief counseling consists in helping the bereaved form a rational belief in life after life. When a person understands that considerable empirical and scientific evidence points to the fact that, in a way which we do not understand, human personality survives the death of the body, the pain of a loss becomes more tolerable. This comes with the realization – not based on faith, but on a rational, critical review of the evidence – that a deceased loved one has not disappeared, vanished into a black nothingness, but rather goes on living in a non-material dimension of existence.

This realization, however, is only possible if, beforehand, we have understood that what we call "ourselves" is in reality our mind – our consciousness, our personality, our being aware, our memories and affections – and not our body. This – our mind – is what goes on living once the body, and the physical brain, have stopped functioning.

But this, unfortunately, is not what the prevailing materialist dogma maintains. Both the media and the academic world are literally dominated by a worldview – materialism – which maintains that all that exists is matter, and therefore mind is merely a product of the electrical and chemical activity of the brain. When that activity stops, mind (that is, ourself) stops.

Well, this may be trumpeted all over popular science media, but is quite simply not true. Not true in the sense that it is NOT what empirical and scientific evidence tells us. And here enters our Everest mountaineer, providing us with an example of just one of the many areas of evidence that disprove the equation mind = brain.

I am myself a passionate mountain climber. I have not climbed Mt Everest, as I like more technically challenging rock and ice climbing (and cannot afford grilling thirty or forty thousand dollars in a Himalayan expedition!), but I have been often enough at high altitude to know well the effects of oxygen deprivation. They are horrible. For me, it's not that much the headache, but rather a sickening and debilitating nausea. The shortness of breath makes moving and concentrating difficult; and increases any sense of fear that may already be present because of the environment.

These effects are already quite apparent at altitudes of about 10,000 feet (3,300 metres) and that is where an Italian professor from the University of Turin - as shown by a BBC documentary I watched recently - investigated a most intriguing application of the well-known placebo effect. He took a number of his students, brought them up to a high-altitude laboratory perched on the Italian Alps, and had them going out for a 30 minute brisk walk on the glacier, wired will all sorts of sensors and monitoring devices. All the subjects were breathing through a mask connected to a gas bottle, and all were told that they were getting supplemental oxygen. But, critically, half of the bottles contained normal air, and not oxygen.

The first remarkable result is that all the participants in the study reported the same subjective reduction in altitude sickness symptoms. Whether they were breathing supplemental oxygen or just air, they did not report headache, nausea or shortness of breath. This could be dismissed as a simple psychological effect.

The professor, however, showed that it was not only the reported symptoms that did not change. All the physiological parameters measured by the instruments were the same for the oxygen group and the control group. This, in itself, is amazing. A pure thought - the belief of breathing oxygen - directly influences physiological processes happening in the body.

But the truly astonishing finding came when the professor looked inside the brain to see what was happening to a particular chemical. When the body is starved of oxygen, the brain produces a natural anti-inflammatory substance called prostaglandin E2, or PGE2. That is what is largely responsible for the reduction in altitude sickness symptoms. Therefore -

read this carefully – a pure thought, an exquisitely "mind" function, makes the brain increase the production of a chemical.

Hold on a sec, Mr Materialist, wasn't mind a mere product of exactly that kind of chemical reaction? How do you explain that "an illusion" (as you define consciousness) directly affects the very processes that you say create it?

If mind did not exist as a separate, independently existing reality, how could it intervene and modify the activity of the brain?

It was indeed a long way home, but our Everest mountaineer helped us once more realize that mind is not the brain, that "we" are not our bodies. When we understand that, it becomes less difficult to accept the idea that human personality survives the death of the body.

Then, we may ask ourselves how on earth does a thought ("I am breathing oxygen") know that it has to intervene on the production of PGE2 to deliver the expected results. This remains a most fascinating, impenetrable mystery.

What's It Like on the Other Side?

Be it with a bereaved person, with somebody who's expecting to die, or with somebody who's just interested in the subject matter, there inevitably comes a point in my conversations about life after life in which the other person asks – What's it like on the other side? Or – How and when do we reincarnate? And – Why do we come into material existence at all? Etcetera.

These are the Big Questions – questions as old as mankind itself. And, who am I to answer them? Nobody. And, in reality, I don't even try. I am not an enlightened being. I did not gain an intuitive knowledge of All There Is To Know. And, least of all do I consider myself a spiritual teacher.

I am a person of reason, and of science, who tries to make sense of the evidence. Based on my studies, there are things about which I am reasonably certain. To the best of my intellectual honesty, I am convinced that human personality survives physical death. That is the most parsimonious explanation for the evidence we have. But I don't know how this happens – my certainties end there. And, especially, I don't have any answer of my own to the Big Questions.

However. Asking the Big Questions is perfectly legitimate. So is speculating on possible answers. As long as we agree that this is an exercise of intellectual exploration and we are not presenting revealed truths, it can be a very interesting activity.

And, in my opinion, the best way to explore an unchartered territory is to review the testimony we have from people who live there. If we agree that, based on the available evidence, the spirit communicators who talk to us through mediums are indeed who they claim to be – discarnate personalities inhabiting a non-material dimension of existence – then their testimony can provide a lot of material for our speculations about the afterlife.

This has been done several times, and by scholars much more learned than me. In his 2009 book *Life Beyond Death*, for instance, the late and sorely missed Prof David Fontana draws on a vast range of sources to attempt a systematization – a "Lonely Planet Guide" to the different levels of the afterlife. In his 2011 *The Afterlife Revealed*, author Michael Tymn goes even further: along with a similar systematization, he provides a bewildering selection of the original quotes from spirit communicators to support his speculations.

Today, however, I would like to talk about the most recent addition to the "afterlife travel guides" – the product of an altogether different and extremely interesting exercise.

From his website, we learn that

> "Jeffrey A. Marks is a spiritual medium and researcher, paranormal investigator, and a dynamic educator and speaker on spiritual potential. Jeffrey is a compassionate voice for the spirits, and as a medium is known for his humble authenticity, by sitters and spirits alike. His groundbreaking book The Afterlife Interviews: Volume I, is the definitive guide to what life is like on the Other Side, and was a finalist in the 2013 USA Best Book Awards in New Age Non-Fiction. Jeffrey's first book, Your Magical Soul: How Science and Psychic Phenomena Paint a New Picture of the Self and Reality, won the 2012 Nautilus Silver Book Award in Science & Cosmology, and was a silver finalist for the 2012 IBPA Benjamin Franklin Award in Body, Mind & Spirit. His writing explores new territory by including a left-brained approach to an examination of right-brained themes of metaphysics, higher mind potential, spirit communication, the nature of the afterlife, and paranormal phenomena – while using the lenses of both science and spirituality to understand the fullness of the human experience."

I was asked by his publisher to review his most recent book which is the continuation and conclusion of the work he initiated with *The Afterlife Interviews Vol. I*. Reading it, I got really excited. What I found most interesting, however, was not the substance, or the conclusions. Don't get me wrong – substance and conclusions are very interesting, but, to my eye, not very different from what I had seen in other publications on the same subject.

What really got me excited was Jeffery's approach to collecting information from the sources. Instead of reviewing quotes from spirits communicating through other mediums, he made use of his own mediumship gift and actually asked, in a systematic way, a selection of questions to 14 discarnate personalities.

Starting with a list of fifty-two questions regarding the nature of life on the Other Side, Jeffrey went directly to the spirits for answers. Using an analytical approach, Jeffrey recorded fourteen mediumship sessions with individuals from various faiths and backgrounds and interviewed their friends and loved ones from the Other Side. The result may well be the most direct and accessible description of the nature of the afterlife compiled in modern times.

In Volume I of *The Afterlife Interviews* series, we hear directly from those on the Other Side about: the nature of the dying process itself; the Life Review; the new spiritual body; new knowledge acquired after passing; how language is facilitated; how spirits exist in the framework of Time; spiritual evolution and levels; reincarnation; the nature of evil; and much, much more.

Volume II of *The Afterlife Interviews* continues where Volume I left off, going deeper to answer The Big Questions. We hear what the spirits have to say about: access to universal knowledge; validity of the Oversoul; survival of animal souls; the afterlife environment, occupations and economics of exchange; cities and homes in the afterlife; making new souls; spirit guides and angels; the role of religion; existence of the prophets; fate of the evil; fate of suicides; and the spirits' views of God.

With this work, Jeffery pursues a novel approach and fills a gap in afterlife studies. His books provide a lot of interesting material to support our speculations. What his "interviewees" say is perfectly in line with what has been recorded from alleged similar sources during the last century and a half, but the level of depth is generally greater. Plus, having fourteen different sources talk about the same specific subjects provides much greater detail and finesse.

A Case for Blind Rage

Rage is a horrible thing. Part emotion, part behavioral reaction, it is the motivator behind some of the ugliest things man is sadly known for.

A particularly vicious form of rage is the "righteous rage" - those situations in which we are absolutely convinced that we are right, that we are defending one position which is, in our opinion, so self-evidently true that we cannot even imagine others can see things differently. And that in itself makes us more angry. "It is his fault that we had an accident. He did not give way at the intersection, even if I had the right of way"!

This kind of rage is what I am battling with these days, and I hate it. To explain why, I'll have to ask you to please read carefully what Wikipedia says about "Electronic Voice Phenomena", which I personally consider one of the many (and one of the strongest) elements of evidence for the survival of human consciousness of bodily death:

> "There are a number of simple scientific explanations that can account for why some listeners to the static on audio devices may believe they hear voices, including radio interference and the tendency of the human brain to recognize patterns in random stimuli. Some recordings may be hoaxes created by frauds or pranksters.
>
> Psychology and perception
>
> Auditory pareidolia is a situation created when the brain incorrectly interprets random patterns as being familiar patterns. In the case of EVP it could result in an observer interpreting random noise on an audio recording as being the familiar sound of a human voice. The

propensity for an apparent voice heard in white noise recordings to be in a language understood well by those researching it, rather than in an unfamiliar language, has been cited as evidence of this, and a broad class of phenomena referred to by author Joe Banks as Rorschach Audio has been described as a global explanation for all manifestations of EVP.

Skeptics such as David Federlein, Chris French, Terence Hines and Michael Shermer say that EVP are usually recorded by raising the "noise floor" – the electrical noise created by all electrical devices – in order to create white noise. When this noise is filtered, it can be made to produce noises which sound like speech. Federlein says that this is no different from using a wah pedal on a guitar, which is a focused sweep filter which moves around the spectrum and creates open vowel sounds. This, according to Federlein, sounds exactly like some EVP. This, in combination with such things as cross modulation of radio stations or faulty ground loops can cause the impression of paranormal voices. The human brain evolved to recognize patterns, and if a person listens to enough noise the brain will detect words, even when there is no intelligent source for them. Expectation also plays an important part in making people believe they are hearing voices in random noise.

Apophenia is related to, but distinct from pareidolia. Apophenia is defined as "the spontaneous finding of connections or meaning in things which are random, unconnected or meaningless", and has been put forward as a possible explanation.

Physics

Interference, for example, is seen in certain EVP recordings, especially those recorded on devices which contain RLC circuitry. These cases represent radio signals of voices or other sounds from broadcast sources. Interference from CB Radio transmissions and wireless baby monitors, or anomalies generated through cross modulation from other electronic devices, are all documented phenomena. It is even possible for circuits to resonate without any internal power source by means of radio reception.

Capture errors are anomalies created by the method used to capture audio signals, such as noise generated through the over-amplification of a signal at the point of recording.

Artifacts created during attempts to boost the clarity of an existing recording might explain some EVP. Methods include re-sampling, frequency isolation, and noise reduction or enhancement, which can cause recordings to take on qualities significantly different from those that were present in the original recording."

So intelligent. So "clinical". So convincing, isn't it?

Now I tell you what I would really like to do. I would like to take one of these sorry souls we call "the skeptics", tie him (or her) to a chair, have him (or her) watch the documentary *Calling Earth*(*) and figuratively kick him (or her) in the butt until he (or she) admits that the "normal explanations" are ludicrous in the face of the evidence we have, and the Wikipedia entry is a pile of horse manure masquerading as "scientific truth".

My blind rage is further fueled by the thought that millions of people – "the public" – are cynically, consciously, voluntarily misled by these self-appointed Teachers of the Truth. We are denied the opportunity to know about, and to make our own judgment on, possibly the most important question there is to ask – What happens when our body dies?

But I will not give in to rage. The guy who does not respect priority on the road, and the skeptics, are there to push me into being somebody I don't want to be. So I suggest that you watch this absolutely fantastic, 60 minute documentary on EVP, and, instead of feeling furious at the intellectual dishonesty of the skeptics, marvel at the amazing wonders and mysteries this life of ours presents us with.

(*) The documentary is found at: http://vimeo.com/101171248

An honors degree in mathematics with no detectable brain?

The predominant view amongst neuroscientists is that human consciousness is a product of the electrical activity of the brain. They call it an "emergent property" of grey matter. The basic assumption is that if you take approximately 100 billion nerve cells and you wire them together through a prodigious network of connection, consciousness somehow "emerges" spontaneously from that complexity. One neuron is not conscious, but 100 billion, taken together, are.

This materialist theory of consciousness is up against some formidable challenges, all happily ignored by the mainstream materialist neuroscientists. First and foremost, there is a fundamental logic obstacle known as the "explanatory gap". This obstacle can be described as follows - Can a motorcycle fly? Obviously not. Can 100 billion of them fly? No, they still can't. So then, how is it that an unconscious piece of matter – a nerve cell – becomes conscious if connected to other equally unconscious pieces of matter?

There are a lot more challenges to the mind = brain equation, and some are so huge, so self-evident that they should force a little reflection on the part of our materialist colleagues. Unfortunately, this doesn't happen much. With this post, I will tell you about one of these formidable challenges. More will come in future posts.

In April 2007, Science, possibly the most reputable scientific journal on the planet, came out with an interesting – to say the least – article. It tells about the remarkable research conducted at the University of Sheffield by neurology professor the late Dr John Lorber.

"There's a young student at this university," says Lorber, "who has an IQ of 126, has gained a first-class honors degree in mathematics, and is socially completely normal. And yet the boy has virtually no brain." The student's physician at the university noticed that the youth had a slightly larger than normal head, and so referred him to Lorber, simply out of interest. "When we did a brain scan on him," Lorber recalls, "we saw that instead of the normal 4.5-centimeter thickness of brain tissue between the ventricles and the cortical surface, there was just a thin layer of mantle measuring a millimeter or so. His cranium is filled mainly with cerebrospinal fluid."

The student was suffering from hydrocephalus, a condition in which the cerebrospinal fluid, instead of circulating around the brain and entering the bloodstream, becomes dammed up inside.

Normally, the condition is fatal in the first months of childhood. Even where an individual survives, he or she is usually seriously handicapped. Somehow, though, the Sheffield student had lived a perfectly normal life and went on to gain an honours degree in mathematics. This case is by no means as rare as it seems.

In 1970, a New Yorker died at the age of 35. He had left school with no academic achievements, but had worked at manual jobs such as building janitor, and was a popular figure in his neighbourhood. Tenants of the building where he worked described him as passing the days performing his routine chores, such as tending the boiler, and reading the tabloid newspapers. When an autopsy was performed to determine the cause of his premature death he, too, was found to have practically no brain at all. Professor Lorber has identified several hundred people who have very small cerebral hemispheres but who appear to be normal intelligent individuals. Some of them he describes as having 'no detectable brain', yet they have scored up to 120 in IQ tests.

No-one knows how people with 'no detectable brain' are able to function at all, let alone to graduate in mathematics. That is, if one assumes that mind functions are solely the product of the physical brain...

Do I believe in life after life?

This article is more a reflective piece and less of the usual information sharing on applied psychical research and consciousness studies – I hope you will find it interesting nevertheless.

The train of thought I am going to share here originated from the latest of the remarkable series of Skeptiko podcast shows. If you don't know about Skeptiko, you really should. Over a period of a few years, retired high-tech entrepreneur and science enthusiast Alex Tsakiris has put together over 240 well-produced shows which amount to no less than an extraordinary journey of discovery. He has interviewed everybody who's anybody in parapsychology and the broader psychical research, notably including near-death experiences, and many innovative thinkers in consciousness studies. He's also had well – and less well – informed skeptics on the show, never failing to expose the ignorance, logical fallacies and – occasionally – outright intellectual dishonesty prevailing in that camp. Do yourself a favour and check out all the shows, available online at www.skeptiko.com.

At the end of one show, Alex, as he frequently does, asked a question to his listeners – What would it take for you to change a belief, an idea or a position you've held for a long time? This question had me reflecting quite a bit on a point which is cardinal to my innovative approach to working with the bereaved and the dying. The crucial question for me is – Do I "believe" in life after life?

At the end of long cogitations, I came back to the answer I formed already some time ago. No. I don't think I "believe" in the survival of human personality of bodily death. At least, not if we take "belief" to mean blind faith in something for which there is no (or little) evidence. A more fitting description for my own position comes from Swiss physicist André Pictet, who, after having thoroughly considered the evidence for survival, famously wrote "I am compelled to believe by the inescapable logic of facts."

Do conclusions based on the analysis of masses of empirical evidence ("facts") amount to a belief? Perhaps, but they certainly do not amount to a faith. For this reason, in communicating with those of my patients who are interested in the subject, and with the public at large, I prefer to use the expression "rational belief", meaning a belief based on rational thought, knowledge and critical examination of facts.

Now – and here Alex's original question comes into play – Do I think I have a bias? More painful cogitations produced a perhaps unsurprising answer. Yes. I think I do have a bias. I find the idea of a "conscious universe" – one in which consciousness is primary and not only the product of the physical brain, and one in which consciousness survives the death of the physical body – infinitely more attractive and interesting than the "all that exists is matter" alternative.

And, again – Do I think that this bias, this preference of mine, affects my critical evaluation of the evidence? This is a particularly sensitive point, as it calls into question my own intellectual honesty. And the answer, to the best of my capacity to look inside myself, is "no". Yes, I do like to find out that the evidence points to consciousness existing independently from the brain. But, no, I am not subject to let this "liking" interfere or to make me ignore evidence that goes against my preferred position.

And, as I observe myself in my on-going reading, studying, exploring, I realize that I am always keenly attentive to any piece of information which may be at odds with my rational belief in life after life. I haven't found one yet, but I honestly think that my healthy skepticism is alive and well.

This, finally, is my answer to Alex Tsakiris' original question. As much as I am "compelled to believe by the inescapable logic of facts", I am absolutely ready and willing to be compelled to believe otherwise, if my critical consideration of the facts were to tell me so.

The ghost in the machine

Fundamental to my cognitive, patient education approach is helping people understand that the much-trumpeted (and instinctive) equation "mind equals brain" does not stand up to critical scrutiny, and, especially, does not account for masses of compelling empirical (facts) and scientific (research in controlled conditions) evidence. When we understand that "we" are not our brain, it is easier to open up to the possibility that "we" survive physical death.

In this article I will discuss a range of most intriguing studies concerning multiple personality disorder, or MPD. MPD is a severe, well-known and much studied condition in which two or more distinct identities, or personality states, are present in—and alternately take control of—an individual. The person also experiences memory loss that is too extensive to be explained by ordinary forgetfulness.

The "alters" or different identities have their own age, sex, or race. Each has his or her own postures, gestures, and distinct way of talking. Sometimes the alters are imaginary people; sometimes they are animals. As each personality reveals itself and controls the individuals' behaviour and thoughts, it's called "switching".

Obviously, the key question is – Are these personalities "real"? Are these simply changes in behaviour, enacted by an unstable personality, or feigned for some obscure psychological reason?

Whilst, lacking a definition of personality which can be falsified, it is impossible to answer the first question, the second one has been investigated by studies and experiments in several domains, and the answer is no. The changes across personalities go far beyond observable behaviour. In fact, apart from marked differences in physiological responses such as allergic reactions or galvanic skin response, or changes in handedness and handwriting, "switching" has been shown to affect some of the very neurological processes that materialists say generate mind (and personality itself...).

Much research, for instance, has gone into assessing sensory changes, particularly visual ones. In three separate studies between 1985 and 1991, researchers found "clinically significant optical differences between alter personalities" on six measures: visual acuity, manifest refraction, colour vision, pupil size, corneal curvature, and intraocular pressure. In a replication study, a control group of nine people was added, who attempted to simulate alter personalities, for comparison with nine MPD patients. An ophthalmologist, blind to which persons were the patients and which the simulators, administered and evaluated the ophthalmological tests. There were 4.5 times more changes among different personalities in MPD patients than in those of the simulating controls.

Even more striking – and more related to the key point I am trying to make – are the studies on differences in electroencephalography (EEG) profiles. Researchers Miller and Triggiano remarked in 1992, "psychophysiological research using evoked potentials has provided some of the most consistent and convincing experimental evidence for the existence of MPD as a clinical entity, as well as for the distinctness of the personality states". In the pioneering study by Ludwig et al. (1972), "The average visual evoked responses for each personality (four) were quite different from each other," and in fact "each personality had its own individual AER type, as if four different people had been tested".

Now – hang on a second. Are we not told, incessantly, that mind is the "product of the electrochemical activity of the brain"? That consciousness is an "emergent property" (you put together 100 billion unconscious neurons, you wire them together in a really complicated manner, and they miraculously become conscious)? Are we not told, by some academically well-placed extremist, that consciousness is an illusion?

And what do we have here? An "altered state of consciousness" (as the researchers define the personality "alters") changes the way brain responds (electro-chemically) to visual stimuli. Not bad for something that "does not exist", isn't it? The same – repeat, the same – physical brain, not "a" brain which is similar to another, but the same physical object, which allegedly produces one consciousness is actually modified in its innermost functioning by a "personality" change.

You have to explain to us how this happens, Mr Reductionist. If mind and the brain are ontologically identical, as you say, how can one affect the other?

We, the less dogmatic, are left in bewilderment and wonderment. Again, we have hard data (science, Mr Skeptic, published in peer reviewed journals) which is simply incompatible with the mind=brain equation. And we are left to marvel at the idea of not one "ghost in the machine", but, in the case of MPD, different "ghosts", which take control of the stupendously complex "psychoneurosocial" being we call man.

Is there any truth in the law of attraction?

Some people find it difficult to believe that thoughts can influence events. Other people, on the contrary, are happy to go along with the seemingly baloney New Age worldview popularized by the book *The Secret* and its purported Law of Attraction. Funny thing is – there is ample laboratory experimental evidence, gathered under controlled conditions, proving that this is in fact the case. Not only do we have examples of "mind over matter" effects, such as influencing the outcomes of rolls of dice or modifying the output of computer-based random number generators, but we also have plenty of research showing what parapsychologists call Direct Mental Influence on Living Systems. We may come back to this interesting subject in a future article, as in this one I would like to focus on something really extraordinary. Perhaps, the people of *The Secret* indeed have it right…

First of all, we have to ask – Can beliefs and attitudes on the part of the participants in an experiment influence its outcome? The answer has been known for 80 years now, and it's a resounding "yes".

In 1942, Gertrude Schmeidler, professor of psychology at City University of New York, set up a questionnaire to explore students' beliefs about psi. She used the term "sheep" to refer to those who were confident about the reality of psi and "goats" for those who doubted its existence or its pertinence in the context of the test. After the questionnaire, she gave the students a classic psi test with ESP cards in which they tried to guess sequences of target cards. Then Schmeidler compared the results of the psi test and those of the questionnaire. The remarkable conclusion was that the "sheep" had a significant deviation above chance, while "goats" were significantly below it.

This difference between believers and disbelievers, known as the "sheep-goat effect," has been confirmed by many other researchers. A meta-analysis by Lawrence (1992), covering 73 experiments by 37 different researchers, clearly confirms that subjects who believe in psi obtain, on average, higher results than those who do not believe in it.

We all tend to select information which confirms our beliefs and avoid that which seems not to fit with them. Selective perception undoubtedly plays a role in our interpretation of apparently paranormal experiences. Skeptics are justified in stating that those who believe firmly in psi will tend to see its occurrence everywhere, even to the point of confusing their own interpretations with the actual events. On the other hand, disbelievers will also tend toward the complementary fallacy, always finding some so-called "rational" explanation for a psi experience, even when it happens to them. But the sheep-goat effect suggests that the differences run deeper than mere interpretation: one's attitudes toward psi do actually affect the likelihood that such phenomena will occur in the first place. The more an individual takes a materialist view of the world, the less chance such phenomena will emerge; the more one is interested in interconnectedness, and open to psi experiences, the more likely the world will "respond" by creating such experiences.

But an even more interesting question is – Can the beliefs and attitudes of the experimenter influence the outcome of an experiment? This is an altogether different issue, as we are talking about an individual's thought having an influence on a complex process involving many people and a number of variables. Astoundingly, the answer is once more in the affirmative.

Excellent examples of research on experimenter effects in psi research can be seen in two recent experiments by Drs Marilyn Schlitz and Richard Wiseman. Dr Schlitz, the director of research at the Institute of Noetic Sciences, designed a rigorous randomized trial evaluating whether or not subjects could detect another person staring at them from a distance (over a closed-circuit television system). The study yielded statistically significant positive results. When her skeptical colleague, British psychologist Richard Wiseman, failed to replicate the results, he invited her to England to repeat the experiment along with him in two separate but equal trials using the same subjects and the same equipment, and once again she got positive results and he got negative ones.

Mainstream psychological research provides another striking example of the experimenter effect. One particular experiment, repeated thousands of times under the same conditions, measures the time it takes for laboratory mice to find a way out of a labyrinth. At some stage, a group of experimenters were told that the mice they were experimenting on were of a particular breed, genetically selected over generations to be faster at finding their way out of the labyrinth. In reality, the researchers were provided with absolutely ordinary mice. The results of the experiment showed a significant reduction, compared to the average, in the time it took the mice to find their way to freedom.

So, you may be slightly put off by the roaring claims and the New Age lingo of *The Secret*, but there is experimental support for the theory that our thoughts indeed do influence reality.

Awareness during general anaesthesia

I had surgery when I was a boy of 10 or perhaps 11. The surgeon who was going to operate on me was my very own dad. I remember full well, to this day, the details of the last moments before the operation, particularly the anaesthetist's surgical hat, which was yellow and not green like everybody else's . Once an IV line had been fixed to my arm and the anaesthetist was ready to knock me out with Pentothal, I remember my dad's voice asking me to count up to 10. I started in a firm voice, proud to show everybody that I wasn't afraid. One. Two. Three. Coma.

In an instant, I went from full waking consciousness to… death. Yes, for many years I thought that this was what death was like. One moment you are alive, the next moment a black curtain has come over you, erasing everything. Whatever you were when alive, whatever made what you were, is gone. All that is left is a black nothingness. And this, I am afraid, is how most people see the death process.

Only, reality may be quite different. It has taken me ten years of passionate study to arrive at my rational belief in life after life. And, as I said many times in my arzicles, key to that belief is the understanding – based on empirical and scientific evidence – that what makes us "alive" (our mind, our consciousness) is related to but independent from the functioning of our physical brain.

How interesting, then, in light of my previous beliefs about death, that I came across information about a medical congress on "States of Awareness in General Anaesthesia", held in 1989 my "adopted hometown" of Glasgow in 1989. And, in particular, I looked at the lecture given by Dr

Bernard Levinson, relating the results of a controlled experiment he had carried out in 1965 (the original study was published by the British Journal of Anaesthesia).

In his lecture, Dr Levinson, a practicing psychiatrist who had previously worked as an anaesthetist, said, "I had been impressed with the work of David Cheek, a gynaecologist from California. Like him, I had encountered patients who recalled traumatic events during their operations. Like him, I was using hypnosis in the course of my psychotherapy and this recall was always during a hypnotic session. And like him, I was curious. Was this a fantasy or did it really happen?"

He then set about creating a controlled study. There were many factors that had to be monitored and as far as possible controlled. He was offered an operation list at a dental school, where the same anaesthetist and surgeon would perform all the operations. From that list, Dr Levinson chose the first patients who were hypnotizable. Patients had to agree to be subjects for the study, and to have a head halter with electroencephalographic (EEG) leads placed on their heads. Patients were told that their brainwaves would be recorded during their operation and hypnosis would be used after the operation to explore their feelings.

During the operation – after exactly the same combination of anaesthetic drugs had been administered to all patients – the EEG was closely monitored. When the signs of deep cerebral depression with electrical inactivity were apparent, the anaesthetist was signaled that the experiment could begin. By that time – Levinson recounts – the poor anaesthetist was in a near state of panic. He had never taken his patients as deep as that.

The experiment consisted of the team simulating a crisis. Reading from a standard script, the anaesthetist would speak in anxious tones to the surgeon saying: "Stop the operation. I don't like the patient's colour. His/her lips are too blue. I'm going to give a little oxygen." At this point he pumped the rebreathing bag for a few moments and finally announced: "There, that's better now. You can carry on with the operation." The entire ward and operating room staff were warned not to discuss any of such events with the patients.

Each of the subjects was seen one month after his/her discharge from the hospital. They were re-hypnotized and the operation was reviewed in detail. The results were nothing but shocking. Of the ten patients, four recalled verbatim everything the anaesthetist said. They knew who was talking, where he was standing, and described the urgency of the moment.

They understood the crisis and described feeling relieved when the operation continued. Four patients had partial recall. They knew who was talking. They could recall a snatch of a word. They became extremely anxious. Some burst out of the trance and were reluctant to be re-hypnotised. One woman said she felt she was spinning. Each time the circle passed the anaesthetist, she could make out a word. Before she could make any real sense out of his words, she spun out again in a circle. She reacted with an alarm reaction in all six channels of the EEG the moment the operating room plunged into silence.

Dr Levinson said: "This preceded the anaesthetist's words since everyone saw me lift my hand in signal. There was a silence - all eyes on the anaesthetist - and then he delivered his urgent speech. This patient responded to the silence and continued her alarm reaction for many minutes after he stopped his monologue. One young motor mechanic said that the Inferior Dental Artery was cut. I asked him how he knew that. He immediately said that the surgeon had just said that to the assistant. (I corroborated this with the surgeon). He was so wrapped up in this event that the awesome suggestion of the anaesthetist was lost in what seemed like a general panic."

The final two patients could not recall any of the events. They were both identical in their response. It was not as though they were re-living the operation, screening the words spoken and then decided there was nothing unusual to report. They both blocked out the entire operation. They could not be persuaded to look into that darkness.

Now – explanations anyone? Before you start conjecturing, let me tell you something else which is important. By looking at the neural pathway of auditory impulses, it is interesting to note that, unlike all other sensory systems, the auditory pathway has an extra relay. To put things simply, this means that even under the thick chemical blanket of anaesthesia, auditory fibers do continue to transmit sound.

But that, for me, is not the point. What you have to explain to me is how those nerve impulses give rise to a fully conscious experience – together with its emotional correlates and the formation of detailed memories – in a state of "deep cerebral depression with electrical inactivity", when all the structures and mechanisms believed to support consciousness are totally shut down.

Reasons for wanting to know

In this article, I will review the main reasons why I think everybody should learn about the empirical and scientific evidence showing that, in a way which we do not yet understand, human personality survives the death of the physical body. And, by extension, the reasons why I feel so angry that this ocean of compelling evidence is ignored by mainstream science and the media (at best confined to the laughable "paranormal investigation" shows on daytime television) and actively suppressed and/or ridiculed by the sorry souls we call the skeptics.

Before we go into the subject matter, let me concisely state the essence of what we call the "survival hypothesis". What we call "ourselves" – our mind, our consciousness, our awareness, our memories, our feelings and affections – does not end with the death of the physical body. In a nutshell, we are not bodies with a consciousness we lose at death – we are consciousness, with a body we lose at death. This is not a matter of faith, or wishful thinking. This is the most economical (and, believe it or not, the least incredible) explanation for a mass of empirical (documented facts) and scientific (laboratory experiments in controlled conditions) evidence, collected for a century and a half by some of the finest minds on the planet, including a few Nobel Prize winners.

So, why should everybody learn about this evidence, and be allowed to make an informed, rational judgment on whether or not our mind and consciousness survive the death of the physical body?

First: knowledge in itself is good. I think that learning, discovering is one of the most exquisite pleasures available to man. And, learning about ideas that go against appearances and common sense is even more satisfying. The earth is flat, isn't it? That's what our senses tell us. Only, it isn't. The sun goes around earth, doesn't it? That's what our common sense tells us. Only, it doesn't. Newton's laws of mechanics seemed to explain the physical world extremely well, until minute, totally insignificant (for our everyday life) anomalies were noticed that led Einstein to formulate the drastically anti-common sense theory of general relativity. Which happens to be true. Similarly, the fact that conscious life ends at death may appear self-evident, but it is not consistent with the evidence we have. Learning about such evidence is a great, refined pleasure.

Second: knowledge about life after life can help those who fear an impending death (their own or a loved one's) and those who are in pain because of a loss. This is the central tenet of my own approach to grief counselling, my entire website is dedicated to this, and I will not expand on it in this article. Just in itself, this would be an excellent reason for everybody to learn about what appears to be an afterlife, and the fact that information about this subject is ignored, suppressed or ridiculed is a scandal.

Third: understanding that death is not the end can help us to live much better lives. I am not a religious person. In particular, I am uncomfortable with anybody telling me how I should behave – what is good and what is bad. That is a moral responsibility that I consider acutely mine. As much as possible, I try to make this judgment call based on information – knowledge. And I believe that if more people would do that, the world would be a better place. In particular, I believe that if more people were not (consciously or unconsciously) anguished about death, extinction, disappearance, there would be a lot less greed, desire for wealth and power, and, ultimately, much less psychological and physical violence.

Fourth: understanding the process of dying makes for a better "early afterlife". Now, this may really sound "out there", but you have to understand where I'm coming from. Through my studies, through a critical examination of the evidence, I became convinced that those who talk to us through gifted mediums (and in many other ways) are indeed who they claim to be – discarnate personalities who have gone on living in a non-material dimension of existence that we call the spirit world, for lack of a better term. Consistently and insistently, these communicators have been telling us that being prepared, "making friends with death" and knowing what to expect can help to avoid the confusion, disorientation and any negative experience "souls" may encounter after separation from the physical body.

In short, I don't want people to believe me. I don't want people to believe anything. First and foremost, I would like people to be interested, really interested in what I think is the most fundamental question that can be asked – What happens when we die? I would also like people not to be afraid of using their own intelligence. Everybody can learn, everybody can think, judge, evaluate. Finally, I really, really would like mainstream science and media not to be dominated by this ridiculous, unsupported, unscientific dogma that the mind is created by the brain, and when the brain dies, the mind and consciousness die with it.

Pseudoscience

When one expresses the view that the mind is not solely the product of the activity of the brain, and – dare one not! – that aspects of human personality survive physical death, one gets different reactions from different skeptical audiences.

Initially, one is brushed away by being placed in the field of religion, spirituality or "folk beliefs". These are very different categories indeed, but for the skeptic they just amount to people believing anything they have been taught to believe, or have chosen to believe, with no regard for truth or reality. For the skeptic, any such believer is not even worth talking to.

But then, when one makes it clear that a rational belief in life after life is based on compelling empirical and scientific evidence, things get very complicated. The extremists in the skeptic camp – and there are many – will foam at the mouth. Red in the face and with their eyes bulging, they will yell and scream and call you names. Since they cannot physically suppress you, they will insult you, ridicule you. They will go as far as re-writing your own biography on Wikipedia in order to discredit you.

The less virulent of them will still resolutely reject your "claims", maintaining that they are not based on science, but rather on "pseudoscience". And this is when things really get interesting.

Let's look at Wikipedia's own definition of pseudoscience.

> "Pseudoscience is a claim, belief or practice which is presented as scientific, but does not adhere to a valid scientific method, lacks supporting evidence or plausibility, cannot be reliably tested, or otherwise lacks scientific status. Pseudoscience is often characterized

by the use of vague, contradictory, exaggerated or unprovable claims, an over-reliance on confirmation rather than rigorous attempts at refutation, a lack of openness to evaluation by other experts, and a general absence of systematic processes to rationally develop theories. A field, practice, or body of knowledge can reasonably be called pseudoscientific when it is presented as consistent with the norms of scientific research, but it demonstrably fails to meet these norms."

And, in a sudden and dramatic reversal of perspective, it is now my turn to foam at the mouth, scream and yell.

I have been a passionate scholar of psychical research for over ten years now, and my views about the mind-brain question and the survival hypothesis are well known to those who follow my blog. These already strongly-held views, however, have taken even deeper roots since I have successfully climbed the intellectual mountain represented by *Irreducible Mind: Towards a Psychology for the 21st Century*. This is one formidable book, written by a group of extremely well qualified academics, with almost 1,000 pages, 150 of which are of references. It is admittedly not an easy read: the sheer quantity of the information provided, as well as the depth and sophistication of the analysis, are daunting. But this is not the point.

The point is that anybody, I mean anybody in their right mind, with even basic reasoning skills, will be forced, based on the information contained in this book, to conclude that the real pseudoscience is the one that gets trumpeted at us from all corners, the one that dominates the academic world, the scientific press and the popular media.

The real pseudoscience is the materialist view that maintains that the mind is solely the product of the activity of the brain, that consciousness is an "epiphenomenon" – many even say it is just an illusion. This, which gets constantly presented to us as a fact, is in reality only a hypothesis. A hypothesis which counters the most fundamental of human experiences (the one of having a consciousness and exercising intentionality) and which is essentially based on correlates. The fact that certain psychological functions are related to the activity of certain parts of the brain: a) is a lot less clear and scientifically established than we are commonly led to believe; and b) explains nothing as to the nature and origin of consciousness. Furthermore, the materialist hypothesis does not and cannot account for mass of empirical (established facts) and scientific (experiments under controlled conditions) evidence. This is – I suspect –

the reason for the materialists to go berserk and try to ignore, ridicule and suppress such evidence.

But we don't need to consider hypnosis, the placebo effect, neuroplasticity, psychic powers or evidence for survival to see the materialist hypothesis run aground. It is evident that this hypothesis lacks a theoretical foundation capable of explaining even the simplest features of human consciousness. They can repeat that mind is "the product of the electro-chemical activity of the brain" until they are blue in the face, but all the tentative explanations put forward so far (and thoroughly discussed in *Irreducible Mind*) fail to explain how the stimulation of the retina receptors and the electrical activity of the visual cortex give rise to the experience of seeing the color blue. They fail to explain how memory works. Even more fundamentally, they even fail to explain how the intention to move my right arm results in the chain of electro-chemical reactions that ultimately result in me reaching for a coffee mug.

This, the dogma of materialism, is the real pseudoscience. We are constantly lied to. We are denied the right to know. We are not allowed to get the full picture and to make our own judgment. I'd be interested to hear from my readers of any suggestion as to how we – the "public" – could promote, or even demand, fairer, more transparent and complete information on the subject.

PSI – Are We "Cherry picking"?

It really is sad, isn't it, that we proponents of a nonmaterialist theory of the mind have to continuously defend ourselves from the attacks of mostly uninformed or intellectually dishonest critics. In this article, I will challenge one of the most "sophisticated" forms of criticism - a criticism which is leveled by skeptics who at least know and accept that there is serious research showing the existence of psychic abilities, and who are looking for a loophole to be able to discount these experimental results. In my challenge, I will have to be a little technical. I hope, in so doing, not to discourage some of my readers, used to more discursive entries.

Before we begin our review, let's just remind ourselves that a successful study is one that shows results that could not be obtained by chance alone. To measure this, scientists use "odds against chance". Significance in a study is reached if the results are unlikely to be due to chance alone with odds against chance of 20 to 1. Parapsychology research generally delivers experimental results with odds against chance of millions, billions and even trillions to one. This led the notorious skeptic Richard Wiseman to have to admit that "By the normal standards of science, extra-sensory perception has been proven".

Proven. Right? No, actually not right, at least for the skeptic camp. Let's see what some of them have to say.

The sophisticated criticism that I am talking about is referred to as "cherry picking". Skeptics maintain that the positive results of parapsychology research are due to the fact that authors only report studies with significant results and leave the non-significant studies unpublished. This is called the file-drawer problem, referring to reports of unsuccessful studies that may be languishing forgotten in the back of researchers' file cabinets. If the size of this hidden file drawer is large, it tends to inflate the estimate of an overall effect. If the researchers publish only the successful studies, we will come to the unavoidable conclusion that all studies are successful. And this may or may not be true. At the same time, an often-cited criticism of combined experimental results (such as in meta-analyses which look collectively at many studies over several years) is that the overall effect may be due to only a few investigators who reported the bulk of the studies, or to a few studies resulting in exceptional results.

To see how this criticism is addressed, we will review the example of the dice throwing experiments investigating mind-over-matter effects. Dr Dean Radin, Senior Scientist at the Institute for Noetic Sciences, performed a meta-analysis on a large number of such studies. The number of studies conducted per investigator ranged from 1 to 21, with the majority of the investigators (64%) reporting one, two or three studies. Dr Radin calculated the overall odds against chance only for those 25 investigators reporting three or fewer studies (totalling 42 studies). The results remained highly significant, with odds against chance greater than a billion to 1. First conclusion: overall success was not due to a few exceptional investigators.

But, maybe the overall head rate was enlarged by the results of a few extreme studies, perhaps because those studies were flawed in some way. To check this, Dr Radin deleted the "outlier" studies by applying a standard trimming procedure. "Outliers" are studies reporting exceptionally large or exceptionally small effects. In his analysis, it was necessary to delete 52 studies (or 35% of the total 148 studies) to produce a homogeneous set of effects. The overall effect observed in the remaining 96 studies still resulted in odds against chance of more than three million to 1. Second conclusion: the experimental effect was independently replicable even when outliers were discarded, meaning that essentially the same effect had been repeatedly observed in 96 studies.

But then, again, perhaps these successful experiments were published more often than non-successful ones, and that's why such large results were obtained. To assess the effect of the file-drawer of non-successful studies, Dr Radin calculated the number of unpublished, unsuccessful studies that would be needed to reduce the observed odds against chance to less than 20 to 1. For these experiments, the file-drawer number was 17,974. That is, for each study included in the meta-analysis, 121 additional, unpublished and unsuccessful studies would have been required to nullify the observed effect. That would have required each of the 52 investigators involved in these experiments to have conducted one unpublished, non-significant study per month, every month, for 28 years.

This is not a reasonable assumption, therefore the third conclusion is that selective reporting, or cherry picking, cannot explain these results.

Similar analyses, with near-identical results, were carried out on meta-analyses for ESP card experiments and, crucially, for the Ganzfield experiments, considered by many as the most solid body of evidence for extra-sensory perception.

What else can be said? Not much. Again, we have seen that the existence of psychic powers rests on very solid scientific ground and that even the more sophisticated criticisms to parapsychology research do not hold up to scrutiny. How long will it be before mainstream science and the media will wake up to this?

Are we conscious despite our brain? (I)

After having reviewed, in recent months, several pieces of evidence which are incompatible with the currently – and disgracefully – fashionable theory that the mind is produced by the brain, I would now like to discuss evidence compatible with a radically different theory. This theory maintains that the mind, rather than being produced by the brain, is actually filtered, or reduced by it.

Before we start, let me make clear that I do not believe that this is actually a theory. Although it can make generic predictions – some of which are actually confirmed by the studies we will discuss in later postings – the "filter hypothesis" lacks a number of features to be able to fully qualify as a theory. First and foremost, it does not explain how this filter function is carried out, at which level and by which structures within the physical brain (although, as we will see, recent experiments may begin to provide answers...). Secondly, it assumes the existence of a "larger reality" so extremely different from the reality we know in our normal waking state that many will find it difficult to accept. Much as I like the simplicity and elegance of the filter theory, and much as I find it interesting and worthy of further exploration, I therefore prefer to consider it, at this stage, a speculation.

The basic idea behind it was well captured by British novelist and intellectual Aldus Huxley in his famous book *The Doors of Perception*. Huxley wrote, "[E]ach one of us is potentially Mind at Large. But in so far as we are animals, our business at all costs is to survive. To make biological survival possible, Mind at Large has to be funnelled through the reducing valve of the brain and nervous system. What comes out at the other end is a measly trickle of the kind of consciousness which will help us to stay alive on the surface of this particular planet."

In clarifying this concept, Larry Dossey MD writes, "Huxley, like Henri Bergson, Ferdinand Schiller, William James, and others before him, believed the brain functions as a filter, normally shutting out perceptions, memories, and thoughts that are not necessary for the survival and reproduction of the organism. Rather than producing consciousness, these observers believed the brain largely eliminates it, diminishing what consciousness is capable of revealing to us. As astrophysicist David Darling says in his book *Soul Search*, we are conscious not because of the brain, but despite it. Frederic W. H. Myers (1843-1901), the British classical scholar, poet, and philosopher, advanced a sophisticated filter theory of brain function that was endorsed by his friend and colleague William James, the Harvard physician and psychologist who is widely considered the founder of American psychology. James, with his superb capacity for metaphor, suggested that the brain acts as a lens or prism that filters, reduces, redirects, or otherwise alters incoming light in a systematic fashion. But James didn't consider lenses or prisms as the ultimate metaphor for the brain. As University of Virginia psychologist and consciousness researcher Edward F. Kelly states in his analysis of Myers' views, "subsequent advocates of transmission or filter models have tended naturally to update this basic picture with reference to emerging technologies such as radio and television that serve as the filter instead of lenses or prisms."

In summary, then, there is a vast, wild world out there, but we only get to see a tiny part of it. We already knew this from physics and from the knowledge of how our sensory systems are built and function. The world we know, for instance, would appear dramatically different if our eyes could see infrared radiation – the heat given off by all objects. If we could only see gamma rays – electromagnetic radiation at much higher frequency than visible light – the world would practically not exist. All we would "see" would be general darkness, with occasional flashes of light, a little like shooting stars on a dark summer sky. We don't see infrared or gamma rays because, for evolutionary reasons, we needed to see the color green of trees, and its nearby frequencies in the narrow spectrum of visible light. All the rest is hidden from consciousness, because it is useless - or even outright dangerous - for our survival as biological organisms. However, the filter theory says, the perceptions we have during certain altered states of consciousness are not simply "created" by the mind. In these conditions, we perceive bits of the vast and wild world out there which are normally filtered out. It is not that we see infrared or gamma rays – it's much more interesting than that. We see at a distance, or in the future, like in the parapsychology experiments. We see dead people and communicate with them. We experience non-physical dimensions of existence, like during near-death experiences. In a similar manner, the theory goes, when we take certain psychoactive drugs we open the filter and we do not create perceptions at random, but rather we access a different reality which is actually out there.

As I said earlier, we do not know which component inside the brain would be responsible for this filtering function. However, we have very intriguing experimental results showing that, when people have unusual experiences, the activity of certain parts of the brain is actually inhibited. That may be the filter malfunctioning. We'll explore this in the next article in the mini-series.

This hypothesis has obvious implications for parapsychology. "PSI-mediated information", for instance, as the scientists call information acquired when all known sensory channels are closed, is a rarity because under normal circumstances it is "filtered out". Only in exceptional circumstances, and in exceptional individuals, the filter broadens and this part of the larger reality becomes accessible.

This hypothesis, however, also has great relevance for my own area of specialty – applied psychical research (the use of the information and knowledge we gain from psychical research for the specific benefit of the bereaved and the dying). In the third and last article of the mini-series, we'll discuss two studies suggesting that the "filter" – whatever that may be – may actually open up as people approach death, giving them a glimpse (and at times a fully blown experience) of the non-physical dimension we call the afterlife.

Are we conscious despite our brain? (II)

The mini-series discusses pieces of evidence compatible with the hypothesis that the physical brain, rather than being responsible for generating consciousness, may actually be acting as a filter, reducing a potentially very wide range of perceptions (and the resulting conscious experiences) to the narrow range that is useful for our survival as biological entities. The study we are reviewing here is phenomenally interesting. It represents a major blow to what I would describe as an "intuitive truth". Please bear with me as I explain.

We are so used to seeing the images produced by functional MRI – scans of the brain, showing in false colors the levels of oxygen utilization in different areas. We are led by popular science articles to believe that this is a good representation of levels of activity. This is not the case: both the proxy (oxygen utilization) and the representation (computer-generated false colors) introduce huge levels of approximation. However, this is not the point. Taking the colors in the fMRI images for what they are – an approximation –, the "intuitive truth" I am referring to implies that a relaxed state of mind (like, for instance, during meditation) should produce an image with little colors, particularly in the frontal areas, commonly believed to be in charge of producing consciousness. Conversely, a highly excited state of mind should produce a kaleidoscope of vivid colors in the same areas. And, what could be more "excited" than during the hallucinations produced by psychedelic drugs? The fMRI image of a person on such a high, one would believe, would be like a painting by a pop artist of the 1970s.

Now, please sharpen your concentration, because you'll have to read a technical abstract from a recent scientific publication.

"Psychedelic drugs have a long history of use in healing ceremonies, but despite renewed interest in their therapeutic potential, we continue to know very little about how they work in the brain. Here we used psilocybin, a classic psychedelic found in magic mushrooms, and a task-free functional MRI (fMRI) protocol designed to capture the transition from normal waking consciousness to the psychedelic state. Arterial spin labeling perfusion and blood-oxygen level-dependent (BOLD) fMRI were used to map cerebral blood flow and changes in venous oxygenation before and after intravenous infusions of placebo and psilocybin. Fifteen healthy volunteers were scanned with arterial spin labelling and a separate 15 with BOLD. As predicted, profound changes in consciousness were observed after psilocybin, but surprisingly, only decreases in cerebral blood flow and BOLD signal were seen, and these were maximal in hub regions, such as the thalamus and anterior and posterior cingulate cortex (ACC and PCC). Decreased activity in the ACC/medial prefrontal cortex (mPFC) was a consistent finding and the magnitude of this decrease predicted the intensity of the subjective effects. Based on these results, a seed-based pharmaco-physiological interaction/functional connectivity analysis was performed using a medial prefrontal seed. Psilocybin caused a significant decrease in the positive coupling between the mPFC and PCC. These results strongly imply that the subjective effects of psychedelic drugs are caused by decreased activity and connectivity in the brain's key connector hubs, enabling a state of unconstrained cognition."

What the heck? *Decreased* activity in the prefrontal cortex? Magnitude of the decrease *proportional to the intensity of the subjective effects*? Does this mean that the more the subject hallucinates, *the less* the brain is active?

Yes sir. It is not me saying this. It is not some self-appointed pop-psi guru saying this. It is a group of 15 top-qualified neuroscientists from universities in the UK and Denmark, publishing in – excuse me – the Proceedings of the National Academy of Sciences of the United States of America.

The concluding sentence of the abstract is unequivocal: "the subjective effects of psychedelic drugs are caused by decreased activity and connectivity in the brain's key connector hubs, enabling a state of unconstrained cognition."

Not only this is compatible with – one would say even supportive of – the brain filter hypothesis. It also suggests something even more radical. The wild world of hallucinations may actually be out there, somewhere, all around us. We don't see it – because the filter is hard at work – but such a world exists.

I don't know what to do with this speculation myself. I find it a truly far-fetched hypothesis, but it is entertained, or at least conceived of, by some very serious people (see, for instance, the book *DMT – The Spirit Molecule* by Rick Strassman MD).

Are we conscious despite our brain? (III)

The first of the two areas of research we consider today is deathbed visions. In 1959, Karlis Osis, psychology professor at the University of Freiburg and Erlendur Haraldsson, psychology professor at the University of Munich, carried out a large study in the U.S. and India. They mailed out detailed questionnaires to 5,000 doctors and 5,000 nurses. The responses they got provided information on over 35,000 observations of dying patients. Their main conclusion was that about 10% of the people are conscious at the moment of their death. Of these, two thirds experience a deathbed vision. Within the two professors' research sample, 1,300 dying patients saw apparitions and almost 900 reported visions of an afterlife.

A number of considerations strongly suggest that these experiences are what they appear to be – a peek into a larger reality, one which is normally hidden by the filter our brain applies to our perceptions. Firstly, deathbed visions are quite unlike hallucinations: they are shorter in duration, more coherent and related to the situation of the patient; they convey a strong sense of reality; they include apparitions of deceased relatives (as opposed to the bizarre characters conjured up by hallucinations); and they bring about serenity and peace, as opposed to the stress of hallucinations. Furthermore, they are not correlated with the consumption of drugs, brain disturbances or a history of hallucinations. Finally, psychological factors such as stress levels or the desire to see somebody are not correlated with the frequency or content of the visions.

One factor, however, is clearly correlated with these experiences: how close the person is to actual death. Studies demonstrate that people who think they are going to die and subsequently don't do not have deathbed visions, whilst people who think they are going to recover and subsequently die *do*. It would therefore appear that it is proximity to actual death that somehow triggers a broadening of the filter. In a way which we do not even begin to understand, as the physical brain heads towards a cessation of its functions, the filter opens up, and the conscious experience of a broader reality (including the perception of the non-physical realm we call the afterlife) is registered by a brain-independent mind.

The second piece of evidence, perfectly consistent with the previous one, comes from researcher J. E. Owens and his collaborators, publishing in 1990 in the prestigious medical journal The Lancet. The title of the publication says it all: Features of near-death experiences in relation to whether or not patients were near death. Here too, we see that "NDEs involving subjectively enhanced cognitive functioning tend to occur more frequently in persons who in fact are closer to death physiologically".

These two pieces of evidence strongly suggest that proximity to physical death causes the mental system to spontaneously begin to operate in a radically different way. Parts of an alleged broader reality become perceptible. We have visions of non-physical dimensions of reality which are independent from a functioning sensory system, and these appear as images in a consciousness which is clearly independent from a functional physical brain. All this certainly fits neatly with the brain filter theory of the mind.

The science of mediumship (I)

I once read a wise note of advice by a renowned science writer: every single equation you put in a book will halve the number of readers. Sure enough, the article which was – by far – the least read in the series of 50 or so I have published up to this point was the one that carried the image of an equation…

I will not give up, however. My aim with these articles is not to be popular (although, it makes me very happy when many people read what I write). At the intersection between psychical research and grief counselling, my aim is to make scientific information available in a format that is accessible and hopefully interesting, and to share with my readers some of my own reflections and considerations. I don't want to fall into the trap of "chasing the audience" – choosing my subjects based on their popularity rather than on the fact that I find them interesting and potentially useful.

Therefore, with this article I am starting another relatively science-heavy mini-series on a critical subject which underpins practically everything my readers and myself are interested in: How do we do science, and how some of the methods employed in natural science apply, in particular, to mediumship research.

The inspiration for this series came from yet another less than positive experience I had during a public demonstration by a medium. The reason for my disappointment is simply that I have come to expect too much. Reading tons of accounts of exceptional séances written by highly intelligent, well qualified and critical observers probably led me to think that each reading would be at the same level. Having viewed and reviewed the video footage of both individual sittings and public demonstrations by some British superstar mediums led me to expect a level of detail, depth, and sophistication in the evidential information provided – and a consistency across time and places – which are obviously the exception rather than the norm. Whatever the reason, the fact is that the dozen or so public demonstrations I attended invariably left me unmoved, if not slightly frustrated.

With so many good mediums doing so much good for so many people all around the world, though, I need a way to soothe my negative – and very subjective – feelings, and reconcile them with reality. The answer, as most often in my life, comes from reason and science.

The scientific method as applied to natural science since the 17th century is based on a series of general principles, which are often mistakenly presented as a fixed series of steps: systematic observation, measurement, experimentation, and the formulation, testing and modification of hypotheses. Not all of these are applied in every scientific enquiry (or to the same degree) and not always in the same order. The history of mediumship research is so long, however, and the quantity of information so vast, that we can certainly find all of the above principles abundantly applied.

There is no doubt about the systematic observation part. Beginning with the investigation of the Fox sisters in the mid-1850's, phenomena linked to mediumhsip have been observed and documented in extraordinary detail for almost 200 years, by some of the finest scientific minds of the planet, including a few Nobel Prize winners. This constitutes a phenomenal basis of empirical evidence – evidence based on the senses of the observers.

But – what if the senses of the observers are wrong? What if their interpretation of what they experience does not correspond to reality? What if they are honest and trustworthy, but they are simply deceived? And, especially, are there normal explanations (i.e. explanations that do not include the survival of personality of bodily death) that can account for what is reported by the observers?

And here enters the second part of the scientific principles: experimentation, and the formulation, testing and modification of hypotheses. In the case of mediumship, what we want to test with experiments is precisely the bottom-line question we asked above: are the thousands upon thousands who have observed and reported mediumship phenomena simply wrong?

In more accurate terms, the basic question is: Can mediums provide *specific* and *accurate* information about a deceased person *in the absence of any communication with the sitter*?

Let's look at the italicized terms, to make sure we understand why they are so important.

"The person was young when he passed" is not a specific statement. What does "young" mean? This statement applies to a lot of different people. Critics say that medium readings are just a collection of general statements, of which a few are necessarily bound to apply to the sitter. With our experiments, we want to test whether this criticism is founded or not.

"The person was 14 when he passed" is a much more specific statement. However, it may or may not be accurate. If the sitter's deceased loved one was 17, for example, the first statement would apply, but it would be too general, whilst the second one, although specific, would not apply. Therefore, to be evidential and answer our research question, the statements provided by mediums must be both specific and accurate.

Finally, the standard criticism of medium readings is that the medium engages in "fishing" (making general statements and progressively making them more specific based on the answers he or she gets from the sitter) and "cold reading" (even if the sitter does not speak, he or she can involuntarily use body language that is read by the medium to infer if a general statement is accurate, and progressively refine it to become more specific). Therefore, in order to eliminate the possibility fishing and cold reading, there must be no communication or interaction whatsoever between the medium and the sitter.

Then – very amusingly, one must say – you see critics making a full 180 degree turn and, in order to explain the results of certain experiments, invoke one thing they would rather die than admit exists: telepathy. What if the medium reads the answers in the mind of the sitter? Apart from the critics' laughable turnaround, this is a fair question to ask. So, the absence of any communication with the sitter required by our research question must be understood to exclude the possibility even of mind-reading on the part of the medium.

So much for the basic research question. We will see in the next few articles how the increasingly sophisticated techniques of mediumship research proved that mediums can indeed provide specific and accurate information about a deceased person in the absence of any communication with the sitter.

This is quite extraordinary in itself, but it is not enough. We still have to demonstrate that the deceased person in fact goes on living in that non-material dimension of existence we call the spirit world. According to many, this hypothesis is not amenable to proof within the methods of science. Amongst them, Dr Raymond Moody makes a solid – if phenomenally complicated – argument that such an answer can only come from philosophy: logic, in particular.

We will stick to science, and we will see how experiments give us strong indications of awareness and intentionality on the part of the deceased, whom we technically define as discarnate personalities. They appear to be aware of what goes on in the lives of the loved ones they have left behind, and to have desires, wishes and intentions, and express them in their communications through mediums. This does not amount to proof, but it is a mighty strong indication!

The science of mediumship (II)

As we tackle some of the basics of the scientific method as applied to mediumship research, it is worth remembering what the basic research question is – Can mediums provide *specific* and *accurate* information about a deceased loved one in the absence of any communication with the sitter? The italicized terms are very important, and if you have not done so already, I suggest you read the explanations I gave in the first article of the series. For the time being, we will stick with this basic research question, and we will discuss the even more interesting stuff (awareness and intentionality on the part of the spirit communicators) at a later stage.

The process of communication between a discarnate personality – via a medium – and the loved ones he or she has left behind is a very complicated affair. If it weren't, we would not be having this discussion here. A number of barriers exist and many variables affect the quality of the communication. Some of these barriers and variables we begin to understand, others we have simply no idea of. These various factors interact with each other, with the result that the quality of a reading (how clear the information is, how understandable, how specific and how accurate) varies across a broad range. Some readings are astonishingly detailed and accurate, others are very generic with only one or two pieces of good information. This is the reason why, in the absence of a clear-cut "black or white" result, and exactly like in many other areas of scientific research, we have to resort to statistics to interpret the findings. Before we dig into experimental protocols, we therefore have to take an elementary-level crash course in statistics as relevant to this particular area.

The first, fundamental concept we have to introduce is "odds against chance". Let's simplify, for the sake of this argument, our research question and ask: Can a medium provide one piece of specific and accurate information? Let's say that we set up our experiment with all necessary precautions and we get one piece of information (the sex of the deceased person, for instance) which is or isn't recognized as accurate by the sitter. Let's also say that we repeat this experiment many times, with different mediums and different sitters, so that we end up with a sufficiently large quantity of data.

If the process was ruled by chance alone, you would expect that quite exactly 50% of the mediums' readings would be recognized as accurate and 50% not accurate. In this case, the mediums simply guess, and if the number of trials is large, they are statistically bound to be correct 50% of the time as half of the population is male and half is female. In statistical terms, this is called the "null hypothesis".

But let's say that, lo and behold, in our experiment the mediums' statements are recognized as accurate 68% of the time. We suspect that something is going on, but to demonstrate an effect we need to understand how unlikely it is that our result (the 68%) could be due to some weird trick of chance. This is done with binomial or multinomial probability tests. Binomial tests apply to the cases (such as in our example) in which two categories are considered. If more categories are considered, multinomial probability tests are used.

These tests produce a decimal number. To interpret that number we need to introduce another concept: statistical significance, or p value. This is just a threshold, commonly set at 5%: if the probability of a given outcome is lower than 5% (0.05 as produced by the binomial test), then an investigator can conclude that the results are not due to chance.

This is not difficult, but let's recap just to make sure we have really grasped the concept. In natural science, in order to refute the "null hypothesis" and be able to say that the results are "significant", not due to chance alone, you need odds against chance of less than 5%, or a p value smaller than 0.05. When that happens, you have shown an "effect". In our example, the effect would be the gift of the mediums in extracting information about deceased persons.

5 per cent means 20 to one. That is significant in science – that is evidence of an effect. Laboratory mediumship experiments have produced positive results with odds against chance in the millions, billions and trillions to one. By the standards normally employed in natural science, a variety of effects linked to alleged medium communication with the deceased have been demonstrated at astounding levels of statistical significance.

Please remember this when we look at experiment designs and results. Before that, however, in the next article I will guide you through a "hands on" experiment that you will carry out yourself based on a real-life medium reading – something interesting and even fun, which will demonstrate another statistical approach called "reverse probability", and may very well shock you.

The science of mediumship (III)

In this article I will guide my readers through a hands-on little experiment using a technique called "reverse probabilities". Those who have taken my video course on evidence for life after life already know this. There is one entire module dedicated to this, showing the filming of a sitting by Scottish medium Gordon Smith made by the BBC, which is then "itemized" and put through the reverse probability procedure in full detail.

We will do the same here, with no video, a little less detail, and based on the record of a different sitting by another celebrity medium. We will explain what itemization means and then you, the readers, will apply the reverse probabilities technique yourself. This should be enough to give everybody an idea of how the technique works and how powerful it is.

One word of caution, however, before we start. Although it uses simple statistics, this methodology is highly subjective. Each person who uses this technique is almost certainly bound to produce different results, and such results can therefore not be compared in a statistically meaningful way. Although it uses numbers, it is essentially a qualitative technique. It is indeed very powerful, however. The results, once analyzed in light of what we said previously about the "significance" of an experiment, can be astonishing.

So, how does the technique work? First of all, the record of a medium sitting must be itemized. This is a process which is applied in a lot of (I would say practically in all) mediumship laboratory research. Anybody familiar with the way mediums share their perceptions knows that readings can be messy. There are a lot of words, only some of which are meaningful, and mannerisms typical to the individual medium. In order to subject the reading to investigation, we therefore need to extract and condense the meaningful information that was communicated into separate short statements. Critically, during the process of itemization, all ambiguous statements are removed. Basically, starting with the full, verbatim transcript of the reading, we want to end up with a list of short statements, each of which can be rated by the sitter for accuracy.

Depending on the sitting, after itemization, one may end up with anything between a few and several dozen itemized statements. For the purpose of our little experiment, let's look at part of an itemized list based on a reading by American medium Alison Dubois during experiments conducted by Prof Gary Schwartz at the University of Arizona. This example is contained in Gary Schwartz's *The Truth About Medium*, and interesting, highly informative and very readable book. During this particular reading, the research medium was completely blind as to both the identity of the deceased person and of the sitter (more on these blinding techniques in the next articles) and the sitter was not even present at the reading. For each statement, we'll briefly look at the sitter's comments.

1. *Deceased grandmother.*

Correct. Susy had called herself my "adopted grandmother".

2. *The woman was short.*

Correct.

3. *The woman was surrounded by flowers, especially roses.*

Correct. She painted flowers, including roses.

4. *She died quickly of some cause associated with the chest area.*

Correct. Susy did die quickly from a heart attack.

5. *She cared about the sitter, but used to tease him.*

Correct. Susy would tease me, sometimes pointedly.

6. *The grandmother is in the presence of a deceased child.*

Correct. Susy had told me that when she passed, her plan was to take care of an infant or a young child who died.

7. *She is showing me a newspaper*.

Correct. Not only had Susy been a newspaper reporter, but the obituary she had written herself in preparation for her passing had been published in the Tucson newspaper that very day.

Now, let's look at what we have here. A series of statements rather typical of what one would get in a medium reading. Although – and this is quite notable – all of them were recognized by the sitter, there is nothing earth-shattering, apart perhaps from the last statement about the newspaper. In comparison, by depth, detail and specificity of the information, the Gordon Smith reading analyzed in my video course is simply astonishing... This is a good example, though, because it helps us to understand how reverse probabilities work in a rather ordinary reading.

Now, it's time for you to exercise some judgment. Go back to the seven statements and ask yourself: How likely would it be that the medium got this piece of information right by simply guessing? In order to do that properly, you have to carefully consider all the possibilities. For certain statements this is easier, for others much more difficult.

For example, statement number 2 is easy. The person could be either short, of normal height or tall. If it was me, I would give this statement one chance in three. Statement number 1 is more difficult: the person could have been male or female, she could have been a grandmother, or a mother, or a sister, any other relation, or unrelated at all. You have to take into account these options, weigh them, and express them as odds against chance the likelihood that the medium could have guessed that exact combination (female and grandmother). One in five? One in ten? One in twenty? Your call.

To do this exercise, I suggest that you write down the seven statements on a piece of paper in order to rate each of them with your odds against chance assessment.

Now it is time to deploy the full power of the reverse probabilities technique. It is actually very simple – you have to sequentially multiply the odds against chance of each statement. To make sure you get this right, I will show you an example based on a fictitious series of five statements with their odds against chances ratings.

Statement number	Odds against chance	Total odds against chance
1	1 in 10 (0.1)	0.1
2	1 in 5 (0.2)	0.02
3	1 in 3 (0.33)	0.0066
4	1 in 20 (0.02)	0.000132
5	1 in 4 (0.25)	0.000033

The value in the last column, 'Total odds against chance', expresses the probability that the medium got all of the five statements correct by simply guessing. In this example, the probability is 33 in 1,000,000 (0.000033). Please remember that statistical significance is reached at 1 in 20, or 0.05.

This shows that even relatively banal statements can add up – if they are all recognized by the sitter – to be very, very significant.

Now's the time for you to do some simple arithmetic. Make sure that you are severely critical in guessing the odds against chance of each of the seven statements. This will only contribute to the credibility of the experiment. At the end, look at your final result and compare it with the standard significance level of 0.05. I bet you'll be surprised.

The science of mediumship (IV)

Let's remind ourselves of our basic research question: "Can mediums provide detailed and accurate information about a discarnate in the absence of any communication with the sitter?"

We are therefore investigating three areas. First - are the statements by the mediums precise, focussed and specific enough to actually mean something to the "sitter" (the person who consults the medium to contact a deceased loved one)? This is the criticism typically levelled by skeptics, who claim that statements by mediums are intentionally vague, so that anybody – especially a bereaved person – can read something into them.

Second – if the statements are indeed specific and meaningful, can mediums deliver them without any previous knowledge about the disincarnates or sitters, in absence of any sensory feedback, and without using fraud or deception? There is no need to say that these are the "weapons" used by the skeptics to try to discredit what we technically define as Anomalous Information Reception (AIR).

Finally – if AIR actually takes place (which in itself I find most extraordinary and fascinating), does this information actually come from disincarnate personalities? Observers of this phenomenon have proposed that such information may be "read" by the medium from the mind of the sitter, or from some sort of "memory" embedded in the fabric of space-time.

To answer these key questions, psychical researchers have employed the most sophisticated investigative techniques since the end of the 19th century. Through exquisitely refined research protocols and with the most rigorous controls, "historical" researchers have undoubtedly answered "yes" to all of the three questions. However, this was "field" research – observing and trying to understand a phenomenon as it happens in its natural environment. It was not until some 20 years ago that mediumship research has been brought into the laboratory, an environment in which investigators have control over most – if not all – of the variables of the process, and can perform quantitative, statistical analysis on the results, the same as happens in any other branch of science.

Experimental protocols address these big questions through what we call "blinding". The best experiments follow a "blind" (single, double, triple, quadruple and even quintuple blind) methodology. A reviewer of one of Julie Beischel's research papers famously said "...but here nobody knows anything about anything!" Exactly, that is precisely the point.

The first "blind" we want to be sure of is that the medium does not know the identity of the discarnate. This is quite easily done. Say that I have a sitting with a medium because I want to hear from my deceased father: the single-blind protocol requires that I say nothing about him. I just sit there and wait for the medium to provide information.

Now, let's assume that at the end of the sitting I find the information the medium has provided both detailed and accurate. I am pretty sure that the medium did not "cold read" me (when the sitter is not allowed to speak, still he or she can involuntarily use body language signs that are read by the medium to understand if a general statement is accurate, and progressively refine it to become more specific) and was not "fishing" (making general statements and progressively make them more specific based on the answers he or she gets from the sitter). But that's me – I am sure. The skeptics are not: they say I am deluded, and I have in fact provided some kind of feedback to help the medium refine the statements. How can we solve this? With the second "blind".

In a double-blind protocol, not only does the medium know nothing about the identity of the discarnate – he or she doesn't know anything about the identity of the sitter either! This is very easily achieved through proxy sitting. In my example, it is not me who sits with a medium to hear from my deceased father - it is, for instance, my wife.

But, hold on a second. The hard-line skeptic will tell you that my wife is bound to know something about my deceased father, and therefore – at least in theory – she could provide feedback to the medium. (You see, by the way, how contorted and unrealistic this hypothesis is? And yet, if there is at least one possibility – no matter how unrealistic – that information could be obtained through "normal" means, the skeptics will invoke that to try to invalidate 150 years of evidence…).

How do you fix that particular problem? Here enters the third "blind": the proxy sitter knows nothing about the identity of the discarnate either.

In a triple-blind protocol, it is not my wife who sits with the medium to hear about my father. It is an experimenter whom I've never even met. The experimenter is given my father's name and his date of birth. He or she communicates that to the medium, and that's it – there is no other information that the medium could possibly obtain by normal means.

Does this work? Spectacularly well, and consistently, since the beginning of the 20th century. In particular, starting from the 1920's, we have published verbatim transcripts of literally thousands of proxy sitting experiments. An analysis of the statements made by the mediums using triple blind protocol shows that these are just as accurate and relevant as the ones provided in a normal setting, when the intended sitter is present. Professor Dodds, the rationalist President of the Society for Psychical Research from 1961-63, supervised a series of proxy sitting tests with the medium Nea Walker and was much impressed. He concluded: "The hypothesis of fraud, rational inference from disclosed facts, telepathy from the actual sitter, and co-incidence cannot either singly or in combination account for the results obtained."

In the next article we'll describe the fourth "blind", and explain how that addresses the specificity question. With real-life examples, we'll see that the information provided by mediums under triple-blind conditions is indeed highly specific for the intended sitter.

The science of mediumship (V)

In this last article of the mini-series on the science of mediumship research we'll discuss the last "blind" introduced in the laboratory procedure to address the last possible criticism of successful medium readings.

This series of articles has followed a clear path. First of all, we have answered the first part of the basic research question: Can mediums provide accurate information about a discarnate? We have briefly explained the technique called reverse probabilities, concluding that – even in the case of mediums' readings which appear of modest quality at first – the probability that a medium provides more than a few correct statements by guessing alone is very low. In the case of the best readings – in which the medium provides ten or more correct, detailed statements – such probability is so infinitesimal that this part of the basic research question is considered as answered, in the positive.

Then we looked at another part of the basic question: Can mediums provide information in the absence of any communication with the sitter? We saw how the knee-jerk criticism by skeptics is unfounded. Mediums cannot either "fish" for information or "cold read" the sitter because the sitter is simply not there. In fact, in the best protocol, the medium and the sitter are separated by three levels of "blinding":

> First blind: the medium doesn't know who the discarnate is. He or she is only given a name and a date of birth (or death).
>
> Second blind: the medium cannot interact with the sitter because another person stands in for the sitter ("proxy sitting").
>
> Third blind: the proxy sitter knows nothing about the discarnate.

Now we have to answer the last part of the basic research question: Is the information provided under such strict conditions not only accurate but also specific? This is very important, because it is possible – at least in theory – that if a medium provides a large quantity of information by simply guessing, at least some of it can be recognized by the sitter. In fact, according to this theory, when there is a lot of information, anybody can find something that fits them.

Furthermore, the critics say, there is a strong possibility of "rater bias". Let's explain. Imagine that, as we said previously, I want to hear about my deceased father. Following the triple blind protocol, an experimenter who does not know either me or my father is given a name and a date, and communicates this to the medium. The experimenter (proxy sitter) records all the information provided by the medium in a series of statements ("itemization"). At the end of the process, I am given the list of itemized statements and I am asked to rate the reading for overall accuracy.

A commonly used rating scale is the following:

> 6: Excellent reading, including strong aspects of communication, and with essentially no incorrect information.
>
> 5: Good reading, with relatively little incorrect information.
>
> 4: Good reading with some incorrect information.
>
> 3: Mixture of correct and incorrect information, but enough correct information to indicate that communication with the deceased occurred.
>
> 2: Some correct information, but not enough to suggest beyond chance that communication occurred.
>
> 1: Little correct information or communication.
>
> 0: No correct information or communication.

Now, put yourself in my shoes. I receive a list of statements, and I know that these were intended for my deceased father. Am I not prone to overrate their accuracy? Most probably yes, hence the problem called "rater bias".

Enter the fourth blind. In our example, at the end of the process I don't receive one itemized list of statements, I receive two. The second one also comes from a real sitting, but it was intended for another person and is introduced as a control. Only one of the two lists is intended for my deceased father, and I am blind as to which is which. If I am able to identify the right list, then we have closed the circle: the information is accurate and specific, in an experimental protocol in which "nobody knows anything about anybody".

What happens in real life, then? One amongst many research papers by Gary Schwartz and Julie Beischel, published in 2007, reports the following results:

> In terms of accuracy, the average summary rating was significantly higher for the intended readings than for the matched controls. It is noteworthy that three mediums produced dramatic findings with summary scores of 5.0 and 5.5; two mediums produced moderate findings (summary scores of 3.5); and none of the mediums produced reversals (i.e., control ratings higher than intended ratings).
>
> When asked to choose which reading was more applicable to them, sitters chose the readings intended for them 81% of the time. Of those 13, seven were rated "clearly more applicable" and three as "moderately more applicable".

In summary, this study confirms and extends the results obtained previously with less sophisticated (single- and double-blind) protocols: there appears to be some sort of anomalous information reception mechanism operating during mediumship readings. Research proves that mediums are capable of providing accurate and specific information about a deceased person when all possible communication channels with the sitter are closed.

Amongst mediums

Recently I was in the U.S. for the second time in just two months. This time, I attended a series of events organized by the Forever Family Foundation, an absolutely fantastic organization that I am so proud and happy to be associated with, and one that I wholeheartedly encourage my readers to join (www.foreverfamilyfoundation.org).

Although in this article I will briefly relate my experiences there, the main focus will still be on mediumship, with some considerations – both technical and human – on this mysterious, unpredictable and most fascinating phenomenon. Incidentally, I apologize to Dr Julie Beischel for "stealing" the title of her most recent book and using it as a title for this article.

The first event I attended was a one-day workshop with bereaved parents, organized by the Foundation to give this "especially special" group of people a chance to learn about afterlife science, to share their own experiences in after-death communication, to discuss the effect of such experiences on grief, and to have evidential communications from their deceased sons and daughters through the help of gifted mediums. Once more, I was in the presence of people who have experienced the very worst life can throw at you, and not only have survived, but are open to learning, and to finding whatever grain of positivity can be found in such a devastating experience. Once more, I was left in awe, and humbled by meeting these champions of humanity.

The workshop then led into a two-day retreat, opening to a larger group of bereaved people and to others who have not suffered a loss but are generally interested in afterlife science and after-death communication. The retreat included formal presentations, more mediumship demonstrations, group discussions, guided meditations and one-on-one healing sessions. The thing that struck me most was the sheer quantity, depth and level of detail of the reports of spontaneous after-death communication experienced by absolutely normal, everyday people. Dazzling apparitions, physical phenomena and the most incredible occurrences of interference with electric/electronic systems seem to be the norm rather than the exception...

Finally, I gave a talk to a large and – I am glad to report – captive audience at Long Island University. This was a "big stage performance" for me, as I am used to the more controlled environment of a university classroom, and I thoroughly enjoyed the experience.

Quite apart from meeting so many marvelous people, what I took home from this week on the East Coast was the satisfaction of having directly experienced, at last, what good mediumship is. Some of my readers will remember that I suffered from a bit of "cognitive dissonance". My studies unequivocally tell me that mediums do indeed talk to the dead, but the experiences I had myself during public demonstrations ranged from lukewarm to outright poor, and this left me a little unsettled.

During the retreat, two in particular of the several Forever Family Foundation-certified mediums who were present gave such excellent performances that I felt I had finally had the direct experience I was looking for. This was a pivotal moment for me, personally, and I treasure the memory.

However, I was yet again reminded of how unpredictable and haphazard the process is. And this made me wonder if perhaps, when I talk about laboratory research, reverse probabilities and fantastic odds about chances, I perhaps convey too much of an optimistic view, which runs the risk of promoting excessive expectations – something I definitely don't want to do.

The process of medium communication is subject to a number of variables, some of which we may know or at least suspect, and many of which we simply have no idea about. The interaction between such variables makes for a quality of reading which is variable not only across mediums, but also for the same medium, across time and situations.

The best way to convey the concept of quality as applicable to this specific context is to refer to a measure widely used in electronics and communication technology: the signal-to-noise ratio (SNR). To do so, let's take an easy example. Let's imagine that I get into my car and, before leaving home, I tune the car radio to a channel where there is no broadcast. All I hear is noise – the harsh hiss coming out from the speakers tells me that a broadcast may actually be there, but the station is too distant, and the signal is too faint and "buried" in noise, below the threshold of detectability. In this situation, the SNR is 0 – all noise, and no signal.

Then, let's imagine that, as I start driving, I get nearer to the station transmitting on that particular frequency. The hiss from the speakers will be occasionally interrupted by fragments of broadcast – a little music, or a few words. The SNR would be, say, 0.1 – 10% signal, and 90% noise.

As I get closer, the periods of good reception become longer. At some stage, the SNR will get to 0.5. And, further on, I will mostly have signal. The music, or the speaker's voice, will be occasionally interrupted by noise. Here, the SNR would be 0.8 or 0.9. Obviously, as I get into the full coverage area of the broadcasting station, I will get a SNR of 1 - all I get is the transmission, and there is no noise.

Does this seem familiar? Can we think of another form of communication which is always there, but may be buried in noise to a varying degree? There you go – this is a perfect description of mediumship, or, rather, of after-death communication as such. The problem here is that no medium, no matter how gifted, will ever have an SNR of 1. At least, this is what research and experience tell me. Different mediums have different "SNR capacities": some will deliver a lot of detailed, accurate and specific information with a few inaccurate and unspecific bits, others will deliver a few accurate and specific bits amidst a lot of "misses" and generalities. And not only that, but the same medium will perform at a certain SNR one day, and at another the next day.

The medium does not know why – nobody really does. The medium has very limited or no control over this, and it is left to us to decide what to do with any information that is being provided. We know by preliminary research that mediumship readings have, in general, a positive effect on the psychological wellbeing of the sitters. But I must be very careful not to convey the impression that I believe that all mediums are completely correct all of the time, or that scientific research supports the idea of mediumship as a replicable, reliable, predictable phenomenon.

Despite our wishes and desires, and despite the best intentions of many gifted and compassionate mediums, this is unfortunately not the case.

Spirit and the process of grief recovery

No matter how many individual stories of loss and bereavement one hears, each new one remains exquisitely individual, personal, and deeply touching. Over the last several months, I have been in regular email contact with one of the users of my video course, a delightful woman named Susan, whom I also met in person during her recent trip to Scotland. In a recent long letter, Susan shared with me the ups and downs which eventually led to the passing of Dan, her "handsome, funny, loving husband of only 5 years". In her letter, Susan also talked about her experiences in dealing with "traditional" grief counselling, and how this not only failed to have any positive impact on the recovery process, but also actively discounted the validity of the spontaneous experiences of after-death communication she believes she had, and which were of considerable comfort to her.

After obtaining her permission, I thought of sharing her experiences with my readers, as hers is a textbook example of the situation many bereaved people find themselves in. In a nutshell, research tells us that, on the one hand, traditional bereavement interventions are useless at best and can at worst be harmful. On the other, non-traditional interventions either implying or entirely based on the existence of an afterlife have repeatedly shown to considerably alleviate grief. Furthermore, thousands of bereaved people report experiences that they understand as communication from their deceased loved ones. Regardless of whether these experiences are "real" or not – and I strongly believe that most are –, it is clear that they have a powerful positive impact on the grief recovery process.

It is essential, I believe, that these experiences are validated – by showing that they are consistent with masses of empirical and scientific evidence for life after life – and given appropriate space in any grief support strategy. Susan's letter is important, because it gives grounding and substance to what I've learnt from books and research. It gives voice to the direct human experience, and it mirrors exactly all I've been hearing from many other bereaved persons, including during the recent Forever Family Foundation retreat I attended in the U.S.

Susan writes:

> "In the weeks that followed my husband's passing, I found myself in worse shape than I could have imagined. Many of the friends and acquaintances I was sure would be there for me were in scarce supply, I had frequent crying jags, felt lost, and could not participate in "normal" conversations about every day matters.
>
> I knew I needed help so I sought out a grief counselor. I don't remember the sessions as being particularly helpful, other than her telling me that my grief reactions were normal, so at least this was reassurance that I wasn't going crazy. I do remember cringing when she reminded me that I was now a "widow", and that I should embrace that word. I don't see how that was supposed to comfort me, and it did not. Then I told her about the following incident that occurred a week after Dan's death:
>
> On that morning at 6:23 a.m., my phone rang once and stopped. Then rang again a few times. I picked it up and it was my friend Jeff. Jeff asked if I was ok and I said yes and asked him why he was calling me so early. He replied that his phone had just rung once and when he looked at the caller ID, it said the call came from my house so he was calling me back. It struck me as odd that both of our phones rang just once at the same time, and that his caller ID would show that I had called him, although I was home alone and sleeping.
>
> The counselor's reply to this incident was that it was a coincidence. A coincidence? I don't see how my friend's phone ringing with my caller ID while I was home alone and asleep could be a coincidence. In fact, I don't see how it's even possible. I pointed this out to her but she just brushed it off with no plausible explanation. I then told her that I was planning on contacting a Medium to see if I could receive further communication from Dan. She said that many widows do this and that it is all fakery and called Mediums "those people". That

is the last time I saw her.

I next tried a counselor associated with Hospice. Dan was in Hospice for the last week of his life and they were extremely helpful and I will be forever grateful. They reached out with counseling options after his death so I made an appointment. All I remember is her holding up a cup and saying something like "This is your grief. You can't go over it or around it. You have to go through it." OK, I thought. Is that it? Then I told her about some signs I was getting from Dan and she said he would always be alive in my memory.

Well, no, *I felt he was alive for real!* So, I did not return. Next stop, hospice group therapy. This group lasted for 8 weeks. Each week had a theme. The first week, everyone told their sad story and we all cried. In subsequent weeks, we talked about our regrets, how we could get through the holidays, how we planned to face the future, etc. Mostly we talked about how much we missed our spouses and how sad we were. I found the group helpful in that it gave me someplace to go every week, but not much beyond that. I felt terrible for the other people in the group, some with much sadder stories than mine, but I didn't see how that was supposed to make me feel better about my own loss. I guess some people are comforted to talk to others going through the same pain, but I did not feel that it helped me in any significant way. The last night of group actually made me angry. The group moderator played a really sad song and everyone cried. Except for me. I was too angry about being manipulated into crying, and I didn't get the point.

At some point during this time, I had made my appointment with a Medium. Although not everything the Medium said was exactly correct, I felt that the he captured Dan's personality and did hit on many things that he could not possibly have known or guessed. The session with the Medium, coupled with the phone call to Jeff's house was starting to give me hope. In time, I stumbled across the Forever Family Foundation and Dr Parisetti's course. Now I felt that I had confirmation that I was not crazy or imagining things.

It is now 14 months since the passing of my husband. I have received several other signs from Dan. After receiving these personal communications, several other readings from Mediums, information from the Forever Family Foundation, the wealth of information from Dr Parisetti's course, and after reading many books on this topic, it is becoming clear that Dan is still with me in some way. This knowledge has helped me to face the future with a bit more hope.

While I still miss his physical presence terribly, knowing that his spirit lives on has given me a tremendous amount of comfort. What saddens me is the lack of knowledge and the closed minded skepticism among those in the helping professions, as well as society in general, regarding the continuation of consciousness after physical death. Not only was I let down by the help that I sought from well-meaning bereavement counsellors and Social Workers, but my pain was compounded by hurtful comments from friends and colleagues when discussing this topic. No matter how happily or gratefully I recounted recent communications from Dan, I was invariably met with less than enthusiastic reactions. These have ranged from a chilling silence to warnings of evil spirits. Let me be clear that these are not mean spirited people. They are good people with good intentions that are genuinely concerned that I am not following the prescribed course of grieving.

I have read statistics indicating that approximately 50% of widows and widowers receive some form of communication from their deceased spouses. Why is it then, that our society labels those that talk about these experiences as deluded and/or unstable? It is precisely these experiences that have helped me to heal and move forward. I am now a volunteer for the Forever Family Foundation, doing whatever I can to try to make a difference. But I realize it is an uphill battle. The popular "ghost" shows and supposed real life dramatizations of evil spirits don't help. For anyone "on the fence" regarding their belief in survival of consciousness, these scary shows terrify them and prevent them from learning more. Wouldn't it be wonderful if these sensationalized TV shows, shamelessly looking for sponsors and ratings, were replaced with educational shows reviewing the latest studies and research in this area? Wouldn't that be more helpful than broadcasting scary ghost stories?

Maybe the healing of our pain due to the physical death of our loved ones has to start with each one of us, in some small way. It is time to speak out about our experiences regardless of our fear of the reactions of others, and to pass down our knowledge to future generations in a matter of fact non-sensationalized manner. We can share the wealth of resources that exists on "life after life", and encourage others to do their own research. I wish I had done so before experiencing the devastating losses in my own life. If we can do this and change the course of grief therapy and general knowledge in our society, perhaps future losses experienced by ourselves and others will be tempered by the comfort of knowing that our loved ones will always be right next to us in an unseen dimension."

Being a medium (I)

After having looked extensively into research supporting the claim - and the experience of thousands around the world - that mediums do actually speak to the dead and that an evidential sitting with a medium can be of considerable benefit for a grieving person, we will now turn to exploring the process of mediumship, as described by mediums themselves. I have enlisted the collaboration of two well respected mediums on the East Coast of the U.S., asking them to answer in their own words four questions which will hopefully help us to gain an insight into this mysterious, fascinating - and maddeningly inconsistent - phenomenon.

My first guest is Janine Baryza-Ly, whom I first met at the Afterlife Awareness Conference in Portland a couple of months ago, and with whom I have kept in regular contact since. In describing her work, Janine says she has been working as a spiritual medium for over thirteen years. She has the ability to communicate with loved ones, angels, guides and many energies in another dimension. Through her gifts she is able to bring healing and the understanding that we really are eternal souls, and that love has no boundaries. It is her soul's passion to share the limitless wonder that the Divine has to offer through her teachings, writings and personal connection to all she meets. Her website is: www.aquestfortruth.com

In this article, we'll hear what Janine had to say about my first question. In the next one, we'll look at her answers to three more "key" questions.

1) Can you briefly describe the indescribable? That is - What happens to you during a reading? What is your subjective experience like? Is the process, as seen from your perspective, more or less the same for every reading, or does it vary?

"Readings are a mystery, even to me who has been working as a medium for 13 years and have had this connection my whole life. The way it specifically works I will honestly say, I am not sure, but I can briefly explain to you my experience and perspective of what happens during a mediumship session. Know that I will do an injustice to the images I see as well as the feelings I go through during this process.

The best way for me to describe contact with the other side, is that there is "more", more feelings than I can express, more images than I have context for and more love than I feel I can transmit. With that being said, for me there seems to be a switch in awareness when I enter into a reading - it is as if what I am normally aware of, meaning everyday observation (the room I am in, time and space), gets put aside and I switch into being aware of not what things look like, but the energy that is around.

The intention and agreement between me and the sitter seem to allow the connection specifically to the person I am reading to come through. A reading can happen in person or over the phone, physical presence does not make any difference within this process. I do not need any names or any objects to help the connection. The discarnates seem to arrive for the person who I am sitting with and always the true message is quite similar "I Love you", "I am OK", "I love you more than you can understand". The more specific information for each person helps the sitter having the certainty that the messages actually come from their loved ones.

Most times before readings I start to "feel" a presence or energy around me. Sometimes I can feel warmth or a cool sensation, sometimes it just literally feels as if someone is standing with me. This presence can be with me in the car, while I am doing house work, etc... And many times myself and the spirit joke around a bit or have a small conversation establishing the contact. They seem to know I am getting ready to talk to their loved one and waiting for the reading to begin. When I start to read, I somehow have a knowing of who I am speaking with first. For example, I will just "know" a person's father has crossed. Once I acknowledge this presence and this connection, then more information comes. It is as if the acknowledgment allows the spirit to know that I am listening and they can then give me more information through that channel. It is at this point where the readings vary a little. Sometimes I get pictures, I will see what someone looks like, or they will show me images that specifically relate to the client. For example, yesterday I had a reading with a woman who lost her father and he kept showing me a detailed picture of this specific house. After describing everything that I saw, she told me I described the house she grew up in. I kept hearing "tell her it is ok", and she started to cry because her father had built that house and they had just sold it. I often "hear" phrases or messages.

When I "hear", it is different than hearing the radio or having a conversation. It is more like what you "hear" when you have a conversation in your head, or when you are talking to yourself, you "hear" but there is no sound. Other times I "feel", and these are the harder sprits to read, because it truly feels as if I am interpreting feelings to messages. It's almost like energetic charades. The spirit gives me feelings, and I try to put those feelings into words. Sometimes these readings come with so much love and emotion, I know I am doing an injustice to the connection and tell the sitter that I know there is more, but please hold my hand and try to feel the love permeating through. I have smelled smoke or perfume, tasted homemade cooking.

The varying ways of connection, I feel, has more to do with the spirit I am communicating with than my receptive abilities. I think it has to do with the way that a soul may connect to another soul. For instance, I love to talk, so I would probably connect to people through words. There are others who don't speak much at all, but connect through their actions, others through their touch, etc.. As we are different here on earth, there are also differences on the other side.

Once I connect to one spirit, then others start to come through. It feels as if they know I can hear/see them and they start to line up, saying "oh she can hear us, lets talk" I can often tell you exactly where I feel the energy coming from, for example; I feel them standing to the right of you or behind you, etc. I normally do not stop reading until I feel the message that I am supposed to bring through is complete.

Another aspect to the reading is that the spirit may not appear to be exactly who the sitter knows them to be. I have often seen the spirit, after the connection was made, change into beautiful light/colors, void of gender, or any physical attributes, as well as any personality quirks. Yet when connecting to the person here on earth, they immediately connect in a way that is familiar to the person.

I often call myself an interpreter because it feels as if I am interpreting a vibration or a frequency in which our souls reside.

The actual process for me is often very similar, yet the way I receive the information can vary from reading to reading. I also cannot control who comes through and who doesn't. I am NOT calling in a spirit, rather I am reading whomever is around."

Being a medium (II)

With this article, we continue the exploration into the process of mediumship, as seen through the eyes of mediums themselves. Janine Baryza-Ly (www.aquestfortruth.com) continues answering my questions on some of the key aspects of a reading. Then, in the next two articles, we will look at how another gifted medium answered the same questions. I am sure this will provide material for very interesting considerations.

We continue with 3 more key questions:

2) What are the factors that, in your experience, affect the quality of a reading (discarnate, sitter, medium, setting, imponderables...).

> There are several factors that affect the quality of the reading. The first one is the ability of the medium to set aside his/her own opinions, perceptions, and judgment. The reading really has nothing to do with the medium - the information and messages are for the sitter, the medium's perspective should be set aside in order to have a clear channel of truth. The next is the sitter's ability to be open and be in truth. I have found the most accurate readings have been when the sitter comes with no expectations one way or the other (whether they are looking for a specific connection or they are skeptics looking for the faults) and is just completely open to what comes through. In these cases the connection seems to be so much more clear. I believe this is because the connection does not have to break through the sitter's preconceived opinions, but can just come through exactly as it is with clarity.
>
> The next is honesty of the reader (the medium). I have found that so many readers claim to know what the other side is like, or what

happens when we die, therefore creating parameters to the connection as well as injecting their own beliefs into the reading. The TRUTH is that we do not know what the afterlife truly holds, we only see glimpses and we hope those glimpses provide enough proof and healing for us here on earth dealing with loss. Once we start to say we "know" I believe we are not allowing the true information to flow. I believe that accepting and admitting our own limitations allows the connection to come through with more clarity.

Another factor is how much pain the sitter is in. I have found it is harder to read those who are in the throes of grief, where they cannot see beyond their pain. It is not that I feel that the connection to the other side is weaker, but my ability to connect seems a bit blocked because of the deep pain that the sitter is in. I often teach that there is a timing involved to receiving a reading. It is personal to each person, but no one should ever get a reading before they feel ready, and we should all honor the process of grief."

3) What role does your psychic gift play in a mediumship reading? That is - Do you get all the information from discarnates, or are you aware of using information coming from other channels?

"I have both the ability to be a psychic and a medium. However, for me it is really easy to distinguish between the two channels. During a reading I will tell you where the information is coming from, whether it be from the discarnate or from my own psychic impressions. With the discarnate it is definitely outside of myself and feels like I am the "medium" or "interpreter" between spirit and the human. The psychic information is me reading the energy that is around and it feels like it comes more from my higher knowing, rather than from something outside myself. During a medium reading I am not using my psychic impressions. I even have had times where my psychic self disagrees with the communicating spirit. In a reading I am as honest as possible and give the sitter all the information that I receive."

4) Do you find that there is a difference between one-on-one readings and group readings?

"Yes of course there is a difference between group readings and on-to-one readings. One-to-one readings are always more intimate, and personal. I feel that the sitter is more open and at ease. There is a certain intimacy that happens with one to one encounters, which is simply impossible in a group setting. Also, think about how we act in a group setting in general, quite apart from a medium reading, and it is different from an intimate, individual setting. I find that in small groups there might be a theme or an overall message for all those attending.

For example, there was an open group (meaning anyone can attend) and that night there was a group of young women in their 20s who had all lost a husband or partner. Not one of the women knew the others, yet they were all brought together that particular night.

When reading in large groups, I often find the purpose being more about demonstrating the connection than being able to give anyone a full reading. When I am in this setting, I am "pulled" towards one particular person or group of persons. There are so many loved ones who have crossed that connecting is a bit different. I look out into the crowd and I see who is "highlighted" - I actually see a person being surrounded by light, and I normally go to that person and then start reading. There is a filtering process, so that you can read just one spirit at a time.

For me, one-on-one the connection is more simple: I know who I am reading and I know what I am reading is connected with that specific person. I know that different mediums feel differently about this topic. However I personally like the intimacy and depth that a one-on-one appointment offers.

The one aspect that everyone should take away is that you are LOVED and you are never alone!"

Being a medium (III)

This is the third of a four-part mini-series in which gifted mediums answer my questions about the process of mediumship as seen through their own eyes. Today, it is my pleasure to introduce Laura Lynne Jackson, whom I had the pleasure to meet during my last visit to the East Coast of the U.S. She greatly impressed me, not only with her mediumistic abilities, but also - and perhaps especially - for her human qualities. One episode in particular stuck in my mind. About two thirds into a four hour (!) group reading during a grief retreat with some 25 participants (an extraordinarily demanding feat in itself, during which she provided what I consider an extraordinary level of evidential detail), at some stage, whilst passing on communication from Spirit to a particular person in the audience, she burst into tears herself, literally overwhelmed by the sense of love she was receiving from the other side. This was no fakery, no stage trick. I know enough about psychology to be able to say that she was really deeply emotionally upset, and it took quite a while for her to recover. I particularly like the fact that she let this happen - she didn't do anything to contain the emotions, or to hide her reactions. She was "emotionally naked", there, in the spotlight, in front of a room full of people. As a human being, I was very touched, and liked this a lot.

Psychic since childhood, Laura Lynne was born with the gift to perceive information from the other side. Fortunately, as both her mother and grandmother also had this ability, rather than being frightened by her abilities, she was encouraged to embrace them. Laura began actively developing her abilities at age eleven and has been reading professionally for over 20 years. Laura sees her work with the other side as one of her purposes this lifetime.

Laura Lynne works with a number of scientific organizations dedicated to exploring and raising awareness about the afterlife. In 2006, Laura Lynne became a Certified Medium by Forever Family Foundation (www.foreverfamilyfoundation.com), an organization dedicated to exploring afterlife science. Laura Lynne also currently serves as a Windbridge Certified Research Medium with the Windbridge Institute for Applied Research in Human Potential where she participates in afterlife research performed by Windbridge Institute Director of Research Dr Julie Beischel (www.windbridge.org). Laura Lynne became certified in 2011 after completing eight thorough screening, testing, and training steps during which her ability to report accurate and specific information about the deceased was scientifically tested under blinded conditions by the Windbridge Institute. Laura additionally works closely with Eternea, an organization whose goal is to help create a better future for Earth and its inhabitants by promoting an understanding of seven postulates concerning the nature of reality, predicated on evidence from contemporary research in science and medicine. This research suggests that some core aspect of consciousness exists beyond the brain, survives bodily death and continues eternally (www.eternea.org).

Laura Lynne Jackson lives in New York with her husband, three kids, a white miniature schnauzer and a tuxedo cat. She conducts private readings as well as group readings and runs spiritual development seminars and classes. For more information, please visit Laura Lynne's website: www.lauralynnejackson.com.

And here she goes answering my questions.

1) Can you briefly describe the indescribable? That is - What happens to you during a reading? What is your subjective experience like? Is the process, as seen from your perspective, more or less the same for every reading, or does it vary?

> "During a reading, a wide screen will appear in my mind. I will see things appear on this screen, as well as hear and at times experience sensations, smells, etc.
>
> The screen is divided into two sections: the left side and the right side. The left side of the screen is where I get all my psychic information from. For example, I see people's auras there (which appear to me in a series of colors) and connect with the sitter's spirit guides on that side of my screen. I also see time lines (similar to historic timelines) drawn mapping the sitters life path with key

events and choices marked on the left side of my screen.

All of my mediumship occurs exclusively on the right side of my screen. Once the psychic wall comes down on the right hand side of my screen and the "door" to the other side opens, the sitters loved ones on the other side will step through and give me information about the sitter and pass on messages.

I begin private readings on the left side of my screen and stay there for about 5 minutes while I wait for the other side to step through on the right hand side of my screen. Once the other side steps through, I tend to stay on the right hand side and listen to the other side. Reading psychically versus mediumistically is a very different process and feels very different to me. Mediumship communication is more of a "receive" while reading psychically is more of a "retrieve."

My readings process stays pretty consistent - except for the fact that when I am doing readings over the phone, my eyes remain shut throughout the entire reading, while for in person readings they stay open. I actually somewhat prefer phone readings as they allow me to fully focus my energy on the other side. When I am reading, there is a sort of time shift that takes place. An hour reading can feel like 5 minutes to me.

Perhaps this is because time doesn't really exist on the other side. I love connecting with the other side because not only do I get to feel the love that passes from the discarnate to the sitter, I also am in a type of soul school when I read, as I listen to the messages from the other side and the clarity they offer about life's meaning and purpose. Often times they will give information about an event that occurred (or one that will occur) for the sitter and in doing so, shed new meaning and light on its purpose.

This in turn helps me to understand that we are always learning here during our time on Earth. Every connection, every person you meet offers you a chance to learn and grow. Sometimes, we fail to fully understand the lessons and growth until the other side helps point it out to us."

Being a medium (IV)

With this article we conclude the mini-series dedicated to the process of mediumship as perceived by the mediums themselves. Laura Lynne Jackson continues answering my questions.

2) What are the factors that, in your experience, affect the quality of a reading (discarnate, sitter, medium, setting, imponderables...).

> I view a reading as a three way triangle of energy. It is built by linking my energy with the sitter's energy and the discarnate's energy. To be honest, I don't fully understand exactly how it works, but I know there is a variance between different readings. I find that readings flow easier when I am at a high comfort level (spiritually alert and not ill or tired, etc). I also feel it can be helpful if the sitter is spiritually "open" so to speak, but I have read for many skeptics and the information usually comes through just as clearly. I also believe that some discarnates energy links more fully with mine for whatever reason, so it is easier for the information to flow more readily. And some discarnates are just excellent communicators! When reading, discarnates often use a different kind of "language" to communicate with me. For example, a discarnate may allow me to "hear" clips of the Italian language (so that I know to say that he/she spoke that language) - but at the same time, the discarnate is giving me information that I fully understand through what I call "thought energy." So no matter what languages were or weren't spoken, they are always able to communicate."

3) What role does your psychic gift play in a mediumship reading? That is - Do you get all the information from discarnates, or are you aware of

using information coming through other channels?

"As discussed earlier, I really don't rely much on my psychic abilities when I read. While I begin individual readings by linking energy psychically with the sitter (on the left side of my screen) while waiting for the other side to step through on the right side of my screen, I usually don't remain on the psychic side of my screen for long. Once the discarnates step through, I stay on the right side of my screen - so the information I receive is from discarnates and I am reading mediumistically. Again, reading psychically feels very different than reading mediumistically: reading psychically is a "retrieving" information while reading mediumistically is "receiving" information. One rule of thumb is that all mediums are psychic, but not all psychics are mediums. I feel that developing one's psychic abilities unlocks, or heightens one's ability to connect with and communicate with the other side. Mediumship communication is more like driving on a highway, while reading psychically is like taking a much slower and smaller road."

4) Do you find that there is a difference between one- on-one readings and group readings?

"Absolutely; it is similar to the experience you might have going out to lunch with one close friend versus a bunch of your friends! However, I love doing both types of readings! During one on one individual readings, discarnates on the other side will usually arrive one at a time, be rather orderly, not interrupt each other and allow each other to take turns. Group readings can be a little different, as different discarnates sometimes jump in, talk over and "steal the mike" from each other (so to speak). It can be harder for me to differentiate at times as I am sometimes abruptly pulled away from one person I am reading and directed elsewhere. Sometimes, many discarnates are vying for my attention to be heard and bring messages to their loved ones, so it can get a little "loud" inside my head so to speak! That said, group readings can be fun and lively - and often times members of the group take away valuable lessons by hearing what another person's discarnate has to say about life and our time here on Earth. After all, we are all more linked and connected than we can imagine. And the other side tells me, we belong to each other."

The Departed Among the Living

In this article, I'm going to provide a brief review of one of those books which should definitely be part of the collection of any person interested in psychical research. The subject – apparitions – is certainly not new, having been extensively covered already well over a century ago by some of the founding fathers of the Society for Psychical Research (*Phantasms of the Living*, by Gurney, Myers and Podmore, 1886) as well as in other investigations and scholarly reviews in more recent years.

What makes this particular book attractive is that it explores the phenomenon in a society – Iceland – in which belief in survival of personality of physical death is very widespread. There have been several surveys of belief in an afterlife in Iceland, the most recent of which (European Values Survey, 2011) reveals that 63.4% of Icelanders declare an acceptance of survival of bodily death, about one third of whom reckon they have had personal experience of it in the form of some kind of contact with the deceased.

What I found really interesting, as an overall observation, is that neither the percentage of people experiencing after death communication nor the content of such experiences are significantly different from countries in which such belief is less widespread. I consider this a further indication that after death communication is a universal human experience, largely unaffected by the cultural context.

The book is *The Departed Among the Living: An Investigative Study of Afterlife Encounters* by Erlendur Haraldsson. The author, Emeritus Professor of Psychology at the University of Iceland, is one of the most experienced field researchers in the psi research community, notably in the areas of reincarnation studies and deathbed visions. Prof Haraldsson brings his trademark rigorous, scholarly methodology to the study of the testimonies collected by him and his colleagues from a total of 449 respondents to various public appeals, and from questionnaires containing as many as 79 questions which the researchers were able to persuade their informants to answer.

Most of the book's 39 chapters are devoted to ways in which the deceased are experienced, such as by sight (the majority), sound, touch, smell, or simply the sense of an invisible presence. Other chapters deal with particular modalities of the deceased's passing, such as suicide, murder or other kinds of violent death. Childhood and widowhood experiences are also discussed in detail, as well as reactions by animals and information received through mediums.

Detailed statistics are of great interest to the researcher who wants to compare how after death communication is experienced in different cultures and situations, but the book has a lot more to offer than that. Its value for the casual reader lies in the extraordinary amount of human testimony, reported basically word by word as it was collected from the participants in the research. There is where I found most inspiration for my own reflections. I found it fascinating to see how this fundamental human experience is filtered through and affected by local variables. The accounts from Iceland are, in essence, exactly the same as you would get from any other country in the world. However, when you look at certain details, you cannot miss the strong influence the environment and lifestyle have.

Iceland is an island nation, mostly populated by seafarers and fishermen. It is therefore no surprise that so many of the encounters are reported to have happened aboard fishing vessels or to have involved fishermen who had died at sea. The striking landscape of this volcanic land also features prominently and many stories reflect the simple, essentially rural lifestyle of most Icelanders during the 1970s and 1980s, when the original research was carried out.

Of special interest are cases in which there was a witness and in which communicated evidence was verified. In a particularly well described one, a woman recalls how she saw an apparition of a patient at the sanatorium where she worked, named Jacob, whom she had invited to visit her and her husband shortly before she died.

I said to him: "Do you promise to come tomorrow?". "Yes, yes, I promise," he said. During the night I woke up and it was like all strength was like taken away from me. I was unable to move. Suddenly I saw the bedroom door open and on the threshold stood Jacob, with his face covered in blood. I looked at this for a good while unable to speak or move. Then this feeling disappeared and I felt as he closed the door behind him. I became my normal self, woke up my husband and told him about the incident, adding "I swear something has happened at the sanatorium". I telephoned in the morning and asked if everything was all right with Jacob. "No," said the nurse, "he committed suicide last night".

The husband told Haraldsson that he remembered the incident well. An interesting detail is that Jacob had died after jumping from a bridge, a post-mortem report noting that there were large cuts on his head. The percipient, who never saw the body, cannot have had any normal knowledge of either the fact of death or the form of it. It is cases such as this – and some are described in even greater detail – that seem to make any alternative explanation to survival unconvincing.

Haraldsson's findings contain some surprises. The percentage of apparitions who had died a violent death (28) is more than three times the national percentage of eight, and is exactly the same as that reported by Gurney et al more than a century earlier. This may lend some credibility to the controversial speculation that those who suffer a sudden, unexpected or traumatic death may remain "stuck" in "lower levels" of the afterlife for some time (please appreciate the inadequacy of the language).

Similarly consistent with what has been reported by other reviews and investigations is the fact that the frequency of apparitions decreases rapidly with the passing of time after death, with an unusual percentage reported during the first 24 hours. Most strikingly – and, again, similarly to what has been reported in any other country where investigations took place – in many cases the apparition was of somebody not known to the percipient, or somebody who was known but not known to have died.

The author's conclusion seems therefore fully justified, "When all the accounts we have collected are considered, it seems impossible to reject them all as deceptions and mistaken perceptions. Something real is there, at least in some of the accounts".

What I believe (I)

This article and the next one are going to be fairly different from the others I have written. In these articles, as my readers will know by now, I tend to share and reflect upon things I *know*. By that I mean things which are supported by empirical and scientific evidence, which I subjected to critical analysis and which I found – to the best of my intellectual capacity and honesty – to be true. These "things" can be summed up in two broad statements: a) the mind cannot be reduced to the activity of the physical brain; and b) in a way which we do not yet understand, significant aspects of the human personality survive the death of the physical body.

Now I would like to briefly talk about things I *believe*.

Although we are going to discuss spirituality and metaphysics, I have not suddenly turned into a new age guru. Yes, I am a teacher. Just like I do in my university classroom, in this article I share with my audience data and information, and I encourage everyone to make a stringent critical evaluation, with the aim of expanding knowledge and understanding. But no, I am certainly not a spiritual teacher. I do not share ideas which I expect to be accepted as a matter of faith, and I am certainly not about imposing (and not even encouraging or suggesting) beliefs.

So, as long as we are clear that the following are my beliefs – not things that I know – and that I share them because I think they may be interesting and stimulate discussion, we can go ahead and tackle some of the ultimate, most fundamental questions man can ask.

What is the universe made of? What is the fundamental ground of reality? Is there a God and, and, if there is one, what kind of God is it? And especially, as the French would ask – à quoi bon tout cela? What is the reason for all this? Why is there anything at all instead of nothing? What is the meaning of the universe, and of life?

So, let's take the plunge.

First, I believe that all is one. I believe that there is one, single, fundamental, undifferentiated reality behind the mesmerising, bewildering variety of the universe as it presents itself to us through our senses. As I did at the end of my book *21 Days into the Afterlife*, I would like to refer to the few words uttered by the Buddha himself, emerging from the altered state in which he attained enlightenment.

In describing the ultimate nature of reality, and his experience of it, the Buddha simply said:

> *Profound calm, free of complexity.*
>
> *Uncompounded luminosity.*

First, let's focus on the first line, and in particular on the words free of complexity, which encapsulate much of the ultimate truth as I see it. To give you a little push, I'll suggest a very simple image, one that helped me greatly in my early days.

Think of looking at a stormy sea: gigantic waves, each with its own distinctive shape, each moving in a different and complicated matter. Waves crashing on the rocks, dissolving into clouds of droplets, and apparently ceasing to exist. Quite a fitting description of complexity, I believe.

Now, think for a moment – what are the waves? The waves are the sea. The waves are just temporary, superficial manifestations of an underlying, more fundamental reality – the sea. When they crash and apparently dissolve, they simply go back to being what they always had been – the sea. A few meters below the surface of what appears to us as the worst storm, there is "profound calm, free of complexity".

So, to restate this part of my beliefs, I would say that the mesmerizing variety of objects that populate the world – grains of sand, rocks, mountains, planets, stars and, obviously, people – are in fact just temporary, superficial manifestations of an underlying, more fundamental reality, and this reality is one.

Appearances (the waves), and reality (the sea). How many times have you heard these concepts already? The Buddhists say that we live in Samsara, the world of illusion: we take the material world that we perceive with our senses for reality, whilst it is not. The Hindus speak of the veil of Maya, a screen that hides from us the real nature of the world.

I believe this because I like it – I find the idea elegant, aesthetically pleasing. I also believe this because, at some fundamental level, it is compatible with the view of reality that emerges from modern physics. And, especially, I believe this because, independently from race, religion, language, spiritual tradition and epoch, people in special circumstances have had a direct experience of this ultimate reality, and have told us in almost exactly the same terms.

By "special circumstances" I mean spiritually transformative experiences (the mystics of all religions), altered states of consciousness (Out-of-Body Experiences), encounters with death (Near-Death Experiences) and… death itself (the spirit communicators talking to us through mediums). Mystics, NDErs, OBErs and spirit communicators all tell us that that the world of creation "emerges" from an underlying reality of a superior order (a "creator", generally referred to as God) just as waves emerge from the sea. Reality is not the waves, distinct from one another. Reality is the sea, undifferentiated, free of complexity.

Another way to express this belief of mine is that the creator and the creatures are one.

Borrowing again from my book *21 Days into the Afterlife*, let's see how this fundamental concept has been expressed with almost identical words by people in totally different circumstances.

One NDEr says: "It became clear to me that all of the higher selves are connected as one being, all humans are connected as one being, we are actually the same being, different aspect of the same being."

A spirit communicator says: "He is the One that appears as many, because He is what He is. He is infinite because He is the One, eternal because He is unchangeable, in reality indivisible because in reality He is the only one to exist. He is complete, because He is the Totality that includes everything."

Another proclaims: "I am a manifestation of the universal force that molds and brings everything into life."

I could go on for quite a while with quotes like those. Instead, let me conclude with a very short review of what the great spiritual traditions of mankind have to tell us. For instance, when we read a passage from the Verse of the Throne that says "Allah! There is no god but He" it doesn't mean say, as it is superficially believed, that there are no other gods, it proclaims that the ultimate reality is one - Allah is all there is.

Similarly, the Jewish Zorah tells us that: "If one contemplates things in mystical meditation, everything is revealed as one", and the Christian idealist Dionysius writes: "It is at once in, around and above the world, super-celestial, super-essential, a sun, a star, fire, water, spirit, dew, cloud, stone, rock, all there is".

Meister Ekhart, the thirteen century Dominican monk, wrote: "In this breaking-through I receive that I and God are one. Then I am what I was, and then I neither diminish nor increase, for I am then an immovable cause that moves all things". From the tenth century Sufi mystic Monsoor al-Halaj comes the pronouncement: "I am the Truth!" and from eighth-century India, Hindu mystic Shankara inspires us by saying: "I am reality without beginning, without equal. I have no part in the illusion of "I" and "you", of "this" and "that". I am Brahman, one without a second, bliss without end, the eternal unchanging truth."

And finally, perhaps the most inspiring of the mystics' quotes, the one I found the most beautiful in literary terms and always makes my eyes water, comes from Moses de Leon, a Jewish Kabbalist and probably the author of the Zorah:

> "God, when he has just decided to launch upon his work of creation is called he. God in the complete unfolding of his Being, Bliss and Love, in which he becomes capable of being perceived by the reason of the heart, is called you. But God, in his supreme manifestation, where the fullness of His Being finds its final expression in the last and all-embracing of his attributes, is called I".

What I believe (II)

The key belief I introduced in the last article, which provides the foundation for all other beliefs I hold, is that the world as we perceive it is in fact a temporary, ever changing manifestation of an underlying, unique, undifferentiated and unchanging reality. This core belief has two essential corollaries.

First, this underlying reality is conscious. Consciousness, awareness on a very fundamental level, is the true ground of reality. In a nutshell, I believe that it is not matter that is primary and responsible for the raising of consciousness. Consciousness is primary, and is responsible for the appearance of matter which – as some interpretations of the findings of modern physics would have it – is some sort of illusion.

Again, this very important belief of mine is grounded in beauty, first. I find it beautiful, elegant, and instinctively attractive. Not being a philosopher, I would not be able to even begin to provide a watertight explanation, a "defence" of this idea. I am simply happy to go along with a view of reality which has been put forward by very different and well-informed sources. I am also intrigued by the fact that this particular view is perfectly consistent with the findings of psychical research, including the existence of psychic powers and the survival of personality of bodily death, whilst the currently fashionable materialist interpretation is definitely not.

You will remember my quoting the Buddha in my last article, and the fact that he described the fundamental ground of reality as "profound calm, free of complexity". You will remember that he also said, crucially, "uncompounded luminosity". From my readings of Buddhist literature, I have understood that the term "luminosity" refers precisely to consciousness, awareness. And I adore the qualifier uncompounded –

consciousness, awareness, is essential, fundamental, as per the Oxford dictionary, "not made up of any other substance". It is interesting to note how "light" as the highest, ultimate reality appears with extraordinary consistency in the accounts by people who have gone through a Near-Death Experience, and how spirit communicators have often described merging into an "original light" as the last step in ascending towards the absolute reality in the spirit world.

Now, the idea of consciousness as the ground of being is constantly presented as "proven according to quantum physics" by many self-appointed New Age gurus, who know absolutely nothing about the subject matter. I am therefore very wary of simplifications based on a few, scattered and poorly understood scientific concepts. However, I cannot help noticing that a number of scientists who have as good a knowledge of the subject as one can possibly have argue for exactly the same interpretation. Instead of going on at length with striking quotes, I invite my readers to look up any writing by Amit Goswami PhD, or John Hagelin PhD, or, for a comprehensive, multi-author review, the book *Mind Before Matter: Visions of a New Science of Consciousness*.

In summary, then, the first corollary to my core belief is that not only the basic, undifferentiated, fundamental ground of reality is conscious – it actually is consciousness.

I hope that the second corollary will not offend any of my readers, as this is certainly not my intention. The problem is that I can find no way to reconcile the idea of consciousness as the ultimate ground of being with the idea of God as presented by any established religion. In short, then, I do not believe in God. Not only do I not believe in an anthropomorphic, personified God, but I do not believe either in any form of God having a specific will, intentions, or plans for the universe. A God interested in and mingling with world affairs would be a reduction, a subset of the ultimate reality, and therefore not the ultimate reality itself. If God is all there is, then it cannot be just a bigger, more powerful version of myself, as essentially portrayed by all doctrines.

And I find it sad that the ultimate truth – the universal metaphysics that underlies all religion, the one directly experienced by the mystics of all spiritual traditions – has been consistently trivialized, cut down, simplified, distorted and adapted for contingent reasons by organized religions, to the point that many of those very same mystical experiencers were brutally persecuted in the name of the prevailing doctrine.

The philosophically-oriented reader may be interested in a profound and difficult but quite marvelous book by Frithjof Schuon, acclaimed as the greatest living authority on comparative religions – *The Transcendent Unity of Religions*, or the more accessible *The Perennial Philosophy* by Aldous Huxley.

Lastly, this second corollary has a possibly even more controversial logical consequence. I am not looking to any holy book to know what I should eat on a Friday, how many times I should pray per day and in which direction I should turn towards when doing so, how I should grow my hair and whether or not I should put salt in my bread. Basically, I do not want to be told how to live my life. Most importantly, I consider the call of what is good and what is not good quintessentially my own responsibility. I try to do good and, especially, try not to do evil because I decided so, because I have chosen what I think is right and better for myself and for others.

The bottom line is that I believe that it is possible to be deeply spiritual – being convinced that there is a greater reality than ourselves, that we are all connected as one – and to live accordingly, without being religious. A few centuries ago I would have been burned at the stake for having said so – today, I just hope I have not alienated too many readers in expressing my views so openly.

Rationalist Spirituality

This is the concluding article of a mini-series in which I discuss my beliefs concerning some of the big questions about "life, the universe, and everything" (from the title of a famous Douglas Adams humour book). Not only are these questions – What is reality made of? Who am I? Is there a God? What's the meaning of existence (mine and the whole universe's)? – as old as mankind itself, they are also inevitably bound to arise when one considers issues such as the survival of personality of bodily death. Once one has established – as I have done, to the best of my intellectual capacity and honesty – that significant aspects of our consciousness and personality go on existing after the physical body has died, one is compelled to go on and tackle the "really big stuff".

But, then, as I have explained previously, I am quickly confronted with a big problem. Whilst questions such as "Is there an afterlife?" can be answered on the basis of empirical evidence, the really, really big questions (God, and meaning, in particular) cannot. They can certainly be the object of rational analysis, but I do not think that the available empirical evidence justifies taking any specific position. Therefore, I stressed that what I am discussing in these articles are my beliefs rather than things I consider to be true.

What is a belief, then, for me? It is something that I like to think of as true. But not just anything – this "thing" I like to think to be true must be compatible – at least in general terms – with a broader framework I am comfortable with, and at least have the potential to explain some empirical evidence, some basic facts of life. This is very, very difficult to explain in a few words. Why, for example, do I believe that the stupendous variety of the universe as we perceive it emerges from one single, unique and undifferentiated source? Because this – at least in general terms – is compatible with the worldview emerging from modern physics – a framework I am comfortable with. Empirical evidence shows that objects are made up of a combination of a limited number of elemental particles and these, in turn, are made up of a few, more fundamental, tiny building blocks. These building blocks appear – I obscenely simplify – to be "localizations" of an underlying field of energy and potentiality, sometimes described in quantum physics as the zero point field. As mentioned previously, we can think of this as tiny wavelets emerging from an underlying sea. We see the wavelets, and think that that is reality, but in fact the reality is the underlying sea.

So, I hear what the mystics, the Near-Death Experiencers and the spirit communicators tell us – "all is one" – I see that this is fundamentally compatible with our understanding of the physical world, and I say to myself: Hey, I don't know for sure if that is true or not, but I like to believe it is.

Intuition certainly plays a role in all of this, and so does beauty. I repeatedly said that part of my intuitively feeling that an idea might actually be true has to do with beauty, fascination. Therefore, as I tackle the last and, for me, most difficult question – meaning – it is about fascination that I want to talk first.

I distinctly remember the first time I heard Dr Bernardo Kastrup explaining the basics of the theory he termed Rationalist Spirituality. He was being interviewed by a podcaster I've been following for years (and about whom I'll be writing shortly) and I remember, as I was driving my car, being literally transfixed. Very shortly into the interview I had this piercing thought: if I were to travel back in time to hear Sir Isaac Newton lecturing on mechanics, I would be in the same kind of awe. The sheer power of the reasoning, the beauty and refinement of the ideas, the laser beam sharpness of the logic literally left me mesmerized, and enthused.

Immediately, then, I bought his book by the same name, *Rationalist Spirituality*, and that simply turned out to be the most important book I've ever read. And, in my life, I've read a book or two.

No, Dr Kastrup is not a guru. And no, I have not been converted to a cult. I have long ago realized that my path, in this life, is not one of *experiencing*, as the mystics do. It is one of learning, thinking, reasoning, understanding, and ultimately knowing. And, lo and behold, once I was done with my thinking on this one book, I strongly felt that I had "arrived" – that I had learned what I had to learn in this life.

This is a pretty big statement to make, but it's true. Again, I do not for a moment claim that what I understood to be the truth actually is the truth. But I like it, and I like it so much that I am content and satisfied to go along with it.

I understand that my readers will have grown impatient by now, and will want me to share my "enlightment", but I do not intend to do so. Although the book is not very long, the quantity and depth of the ideas put forward in *Spiritualist Rationality* is such that it would be ludicrous for me to provide a few paragraphs of synopsis. I can only strongly, enthusiastically encourage my readers to buy the book and prepare themselves for a breathtaking ride.

One caveat – the book is not simple. Dr Kastrup, who has a PhD in computer science and has worked for several top level research institutions including CERN, the European Nuclear Research Centre in Switzerland, calls himself a "hobbyist philosopher". His metaphysical ideas and language, however, are certainly not an amateur's. I've found his book to be quite precisely at the limit of my own capacity to understand: one grain more, and I couldn't have done it. But I did it, and that was part of my fascination. I consider the process of understanding one of the strongest, and more sophisticated, pleasures in life. I am not ashamed to admit that I had several "intellectual orgasms" whilst going through this book!

A second caveat: I, for one, have found in this book a beautiful formulation of the basic meaning of existence which resonates perfectly with all I've heard, again, from mystics, NDErs and spirit communicators. It's powerful, logically compelling, and most fascinating. But this won't tell you why bad things happen to good people. This may provide you with a very deep understanding, but it is unlikely to help anybody coping with the difficulties life throws at us.

Finding that particular kind of personal, exquisitely human level of meaning remains, I believe, a painful part of each individual's path of learning and discovery.

An extraordinary night of mediumship

In this article, I am going to share with my readers my impressions – my utter amazement, I would say – following the public demonstration by Scottish medium Gordon Smith, which I had the pleasure of attending. Those who have been following me for a while will remember that, since I started studying the subject of survival of personality of bodily death and until recently, I had been quite underwhelmed by the mediumship demonstrations I had attended. I know from my studies that mediumship is real, but my own experiences failed to provide me with the "fire" of an overwhelmingly convincing direct experience.

That situation started to change in June this year, as I took part in a series of events organized by the Forever Family Foundation (FFF) on the East Coast of the U.S. There, I could witness a group of FFF-certified mediums in action. Two in particular made quite an impression on me. But I remained wanting for a truly extraordinary experience.

That undoubtedly arrived as I went, together with my wife and a friend, to see my all-time favorite medium, Gordon Smith, at a major theatre in my adopted hometown of Glasgow. Why an all-time favorite, amongst the many very gifted mediums who actively practice mediumship today? I don't know – just a deep, personal liking, I suppose. I like Gordon because he exudes humanity, kindness, compassion, but this is true for many of the top mediums. In his case, though, I particularly appreciate the contrast between his very difficult upbringing – growing up in a deprived and violent area of Glasgow – and the kind of person he turned out to be. I like the fact that, until recently, he kept working as a barber (he is known in the UK as the "psychic barber"), and he never, ever charged money for one-on-one readings (something he does for a small number of sitters who are utterly devastated by a loss). He now makes a living through his books, workshops and through public demonstrations like the one I attended. I also like the fact that he has offered himself as a human subject in mediumship research (such as the PRISM – Psychical Research Involving Selected Mediums – a project sponsored by Professors Archie Roy and Tricia Robertson at Glasgow University).

On the night of the demonstration, the Pavillion theater was packed almost to capacity – I reckon between 350 and 400 people. I was intrigued by the gender imbalance, as perhaps 85 percent of the audience was female. In order to appreciate the very special atmosphere in the theatre, you should understand the Glasgow way of social interaction. This is a city in which everybody will talk to you even if you are a complete stranger, and crack a joke, and laugh away with the special, witty kind of humour that generations of Glaswegians have developed as a defence against the gloom of the weather. So, within minutes, the mediumship demonstration was already and surprisingly looking like a stand-up comedy act, in full Glasgow patter, and with the public gingerly talking back to the celebrity on stage.

But make no mistake. The quality and quantity of the information flowing from Spirit was simply stunning. We are used, with good mediums, to have good "hits" here and there – initials, perhaps a full name, a significant date, or a particularly evidential detail. With Gordon, it was a continuous fire of full names, ages, specific places and addresses in town, events, and circumstances. There was simply no respite.

Amongst the many striking examples, I remember one in particular. At some stage, Gordon moved towards the left of the stage and said "I have a gentleman who has been shot in the head". That's quite a precise statement to make. If you wanted to "fish" for information and wanted to use a violent death, you would have mentioned a stabbing, as some areas of the city are still stained by a tragic knife culture. But he talked about being shot in the head, and went on saying that it wasn't the person's fault – he just happened to be in the wrong place at the wrong time. As he continued giving details, a lady in the audience, a few rows in front of the part of the stage he had moved to, said "I know the person you're talking about – I know the full story".

Gordon then said that there had been no justice for this crime, that the legal proceedings were left hanging. The lady confirmed that. Then he said that the murdered gentleman's mother was still here, and that she took the events so badly that she had to be hospitalized in a psychiatric institution. The lady in the audience said "Yes, I know, I went to visit her in hospital the other day". That was the moment when another lady, up in the gallery, where we were, and in fact a couple of seats to my left, stood up and called "Gordon, Gordon, it's my brother you're talking about. The lady in hospital is my mum."

It is the late Archie Roy, Emeritus Professor of Astronomy at Glasgow University and the one who extensively investigated Gordon Smith who wrote "There is, of course, a wide spectrum of mediumistic ability, from marvelous to mediocre. It would appear that, as with almost any other human activity, there are superstars, stars, and barely luminous glow worms! Sometimes, as has been known for more than a century, the entire range of brightness can be shown by the same medium at different times."

We remain in awe of this mysterious process, one which leaves us with many more questions than answers. And we can only rejoice in the presence of bright, compassionate stars like Gordon Smith, who never seem to have an off day!

Another case for blind rage

This is the second article I publish under a very similar title. I certainly don't want to give my readers the impression that I am prone to rage or I am in any way violent. I am prone to frustration, though, and I can't stand everybody's intelligence and common sense consistently being taken for a ride.

The source of my frustration is, again, our friends the skeptics. In the previous article under the same title I invited my readers to watch a fantastic, in-depth and quite astonishing documentary on the state of the art in Instrumental Trans-Communication (the communication between the world of the living and the world of the so-called dead) through a variety of electronic means. And I was wondering how on earth anybody in his/her right frame of mind could deny that something extraordinary is going on. I was marvelling – and figuratively fuming – at the ignorance (or outright intellectual dishonesty) of the skeptics, who keep claiming that there is not an ounce of evidence for survival of personality to bodily death and, more in general, that psychical research as a whole is "a hundred years of nothing".

Today's bottom line is the same, but the angle is completely different. We are talking here about Michael Brant Shermer, an American science writer and historian of science, best known as an extremist skeptic. Shermer is the founder of The Skeptics Society, and Editor in Chief of its magazine Skeptic, which is largely devoted to investigating what they describe as "pseudoscientific and supernatural claims". The Skeptics Society currently has over 55,000 members. Shermer also engages in debates on topics pertaining to pseudoscience and religion in which he emphasizes "scientific scepticism". Need I say anything more?

Now, if Mr Shermer and his many acolytes were an obscure, marginal group, with no influence on the broader reality, I would have no problem. Everybody, of course, is entitled to their positions and ideas. The problem is that this zealot and obdurate skeptic provides a perfect example of how this dogmatic, ignorant or intellectually dishonest group of people literally dominates mainstream science and – what is much worse – ends up deciding what you and I (the public) get to know, as well as influencing what we think.

It turns out that Michael Shermer is also the producer and co-host of the 13-hour Fox Family television series Exploring the Unknown. Since April 2001, he has been a monthly columnist for Scientific American magazine with his Skeptic column. He is also a scientific advisor to the American Council on Science and Health. And me, poor idiot, who thought that science was a method, and not a set of dogmatic, preconceived ideas...

But – why am I focussing on Michael Shermer today? Because his September 16, 2014, column on Scientific American had an intriguing title: Anomalous Events That Can Shake One's Skepticism to the Core. In the subtitle, he candidly admits "I just witnessed an event so mysterious that it shook my scepticism".

Let's see what our extremist skeptic had to say:

> "Often I am asked if I have ever encountered something that I could not explain. What my interlocutors have in mind are not bewildering enigmas such as consciousness or U.S. foreign policy but anomalous and mystifying events that suggest the existence of the paranormal or supernatural. My answer is: yes, now I have.
>
> The event took place on June 25, 2014. On that day I married Jennifer Graf, from Köln, Germany. She had been raised by her mom; her grandfather, Walter, was the closest father figure she had growing up, but he died when she was 16. In shipping her belongings to my home before the wedding, most of the boxes were damaged and several precious heirlooms lost, including her grandfather's binoculars. His 1978 Philips 070 transistor radio arrived safely, so I set out to bring it back to life after decades of muteness. I put in new batteries and opened it up to see if there were any loose connections to solder. I even tried "percussive maintenance," said to work on such devices—smacking it sharply against a hard surface. Silence. We gave up and put it at the back of a desk drawer in our bedroom.

Three months later, after affixing the necessary signatures to our marriage license at the Beverly Hills courthouse, we returned home, and in the presence of my family said our vows and exchanged rings. Being 9,000 kilometers from family, friends and home, Jennifer was feeling amiss and lonely. She wished her grandfather were there to give her away. She whispered that she wanted to say something to me alone, so we excused ourselves to the back of the house where we could hear music playing in the bedroom. We don't have a music system there, so we searched for laptops and iPhones and even opened the back door to check if the neighbors were playing music. We followed the sound to the printer on the desk, wondering—absurdly—if this combined printer/scanner/fax machine also included a radio. Nope.

At that moment Jennifer shot me a look I haven't seen since the supernatural thriller The Exorcist startled audiences. "That can't be what I think it is, can it?" she said. She opened the desk drawer and pulled out her grandfather's transistor radio, out of which a romantic love song wafted. We sat in stunned silence for minutes. "My grandfather is here with us," Jennifer said, tearfully. "I'm not alone."

Shortly thereafter we returned to our guests with the radio playing as I recounted the backstory. My daughter, Devin, who came out of her bedroom just before the ceremony began, added, "I heard the music coming from your room just as you were about to start." The odd thing is that we were there getting ready just minutes before that time, sans music.

Later that night we fell asleep to the sound of classical music emanating from Walter's radio. Fittingly, it stopped working the next day and has remained silent ever since.

What does this mean? Had it happened to someone else I might suggest a chance electrical anomaly and the law of large numbers as an explanation—with billions of people having billions of experiences every day, there's bound to be a handful of extremely unlikely events that stand out in their timing and meaning. In any case, such anecdotes do not constitute scientific evidence that the dead survive or that they can communicate with us via electronic equipment.

Jennifer is as skeptical as I am when it comes to paranormal and supernatural phenomena. Yet the eerie conjunction of these deeply evocative events gave her the distinct feeling that her grandfather was there and that the music was his gift of approval. I have to admit, it rocked me back on my heels and shook my skepticism to its core as well. I savored the experience more than the explanation.

The emotional interpretations of such anomalous events grant them significance regardless of their causal account. And if we are to take seriously the scientific credo to keep an open mind and remain agnostic when the evidence is indecisive or the riddle unsolved, we should not shut the doors of perception when they may be opened to us to marvel in the mysterious."

Yes, Mr Shermer, this is indeed a nice story. It is absolutely typical, and highly representative of the kind of stories that millions of people around the world have been telling for centuries. The very people you have called gullible, suggestible, unreliable – basically, brainless idiots who are not even worth talking to. Your story, Mr Shermer, is an anecdote, and anecdote, in the world you belong to, is a swear word. When we – the open minded scholars and researchers – insist that anecdotes can help us understand the world and that the plural of anecdotes is data, you and your friends insult us, discredit us, ridicule us. Read the entry about myself on RationalWiki, to refresh your memory.

But your story is different, isn't it? Because it's yours. And that shook you, regardless of the fact that it was in fact a rather minor episode, compared to things that happen to many of the people you think are so easily deceived. You ignored and discounted those stories all your life, and one minor episode was enough to shake you.

This, I believe, could be excusable. What I believe is absolutely inexcusable is your wilful ignorance and active suppression of the colossal quantity of evidence produced for a century and a half by some of the finest scientific minds of the planet. Anybody – PhDs, university professors and Nobel Prize winners – who has been at a pain to investigate in depth such body of evidence ended up convinced that aspects of human personality survive physical death.

But you believe otherwise, and based on that belief, you decide what "science" should deal with. And what the public should hear. You decide that the bereaved should be deprived of the comfort of knowing that their deceased loved one goes on living in a non-material dimension of existence. By suppressing and ridiculing psychical research and researchers, you deprive all of us of the possibility of making an informed decision. That, Mr Shermer, is inexcusable.

But people change. I understand from your biography that, before converting to scientism and fundamentalist materialism, you used to be a fundamentalist Christian. I hope that your "episode" will be enough to encourage you to look other people's experiences and at the body of research with different eyes. I do not hope you will convert to anything else. I just hope you will use your considerable brain to look at the facts for what they are, and draw conclusions from them. Have the courage to go where the evidence takes you, despite the prevailing dogmas. And, especially, I hope that you will stop preventing us from doing the same.

Awareness during resuscitation

In the Fall of 2014, mainstream media was flooded with the news of the publication of some of the much-awaited results of the AWARE (AWAreness during REsuscitation) prospective study, carried out by a large team of researchers between in the UK and the US and published in the prestigious, peer-reviewed medical journal Resuscitation.

The very fact that that an apparently obscure piece of scientific research – the conclusions of which are in sharp contrast with the prevailing materialist dogma – is picked up by news organisations and given, for a few days, worldwide prominence should make us happy. And it does. Together with happiness, however, I feel a lingering sense of frustration. As usual, the way in which media deals with stories is dishearteningly superficial – there is no background, no history, no context. It's just the last piece of information that counts.

The Independent, a much respected UK daily, carried the news of the study under the impressive headline "Life after death? Largest-ever study provides evidence that 'out of body' and 'near-death' experiences may be real". The reader has the impression that, finally, some "serious" research has been done on the "Hollywood movie" subject of Near-Death Experiences. As reported in the article, the results are indeed intriguing. But... What about the rest of the "serious" studies carried out during the last 30 years by an impressive array of researchers?

In terms of phenomenology of the experience, what does the AWARE study have to add, for instance, to the 2001 prospective study published by Pim Van Lommel and others in The Lancet? In terms of veridical experiences, what about the two major rounds of investigations by cardiologist Dr Michael Sabom, or the PhD dissertation of Penny Sartory in the UK? And these are just a couple of examples.

The bottom line is: the Handbook of Near-Death Experience (the "bible" of NDE research edited by Bruce Greyson and others) cites no less than 65 scientific articles published in peer-reviewed journals – why is this impressive body of research completely forgotten, and the AWARE study presented as something new? The public should know that this is not an isolated piece of research, and that its findings corroborate and confirm what NDE researchers have been telling us for over 30 years.

And yet, not only the AWARE study itself is very important but also, notwithstanding my own grumpiness, the fact that it has been given so much visibility is indeed very positive.[1]

Now, from the substance point of view, the study contains important elements which I found at odds with my own understanding of "episodes of consciousness happening in a period in which cerebral activity is severely impaired or non-existent" (as, typically, during cardiac arrest). These, I believed, basically consist in a large majority of pleasant experiences fitting to varying degrees the "NDE model", and a minority of negative experiences described as "harrowing NDEs".

The AWARE study tells us otherwise. Even the background of the study is different, as it was not primarily undertaken to study NDEs, but rather to investigate the reasons why cardiac arrest survivors are known to experience "cognitive deficits including post-traumatic stress disorder, depression and memory loss". Could this be the result of traumatic conscious experiences during periods in which no consciousness is expected? If so, what about the beneficial, positive psychological and behavioural changes Near-Death Experiencers show, even decades after their episode?

And, lo and behold, the researchers reported that, amongst the 140 subjects of the AWARE study, 46% had memories with 7 major cognitive themes: fear; animal/plants; bright light; violence/persecution; deja-vu; family and recalling events post cardiac arrest. This is a very, very different picture from the one I had about NDEs. Much more varied, diverse, much less "rosy", not spiritual at all, and not particularly desirable.

---

[1] The original article is available at http://www.horizonresearch.org/Uploads/Journal_Resuscitation__2_.pdf and makes for highly recommended reading by scientifically-inclined readers.

And, yes, 9% of the sample had experiences that could be classified as "typical" NDEs, while 2% described awareness with explicit recall of "seeing" and "hearing" actual events related to their resuscitation.

Much as I remain fascinated by the fact that conscious experiences can be had when there is no brain, the variety of such experiences leaves me puzzled, and confused. I knew that only about 10% of cardiac arrest survivors have a full NDE – a finding which has been pretty constant across several studies – and always wondered why. What happens to the rest? Do they have an experience, and don't recall it, as proposed by some prominent researchers? Is it only those who are closer to the point of no return who have an experience, as proposed by others?

But now we learn that close to half the cardiac arrest survivors have experiences that, on the face of it, look more like dreams than anything else. So the questions flood my mind. Why do some people "dream", and others perceive veridical details about their procedures and the surroundings? Why do some people have what looks like bad dreams and other have marvellous, life-transforming Near-Death Experiences?

As ever in psychical research, our logic and common sense are challenged. Science – as a method and not a set of dogmatic beliefs – forces us to follow the data and go where the data take us. Data from psychical research and consciousness studies take us to some very strange places indeed, and we can never rest. Our capacity to understand, to fit the data into some kind of explanatory framework is constantly stretched. I'd love to hear from my readers if I was the only one to feel confused upon learning about the AWARE study.

One essential acknowledgment

This book would have never seen the light of the day without the invaluable help of my two editing angels: Susan Rizzi, a dear friend from New York, and my own lovely wife Angela Higney. I owe both of them immense gratitude.

Is there an afterlife? Does human personality survive bodily death? Scientific research and a colossal amount of compelling empirical evidence, gathered for over 150 years by some of the finest minds on the planet, seem to indicate that Clint Eastwood's 2011 movie Hereafter actually got it right. Consciousness indeed appears to exist independently of a functioning brain, and to extend well beyond the transition we call death. Written by a medical doctor, university lecturer and member of the Society for Psychical Research, 21 Days into the Afterlife makes the case for the survival hypothesis in a series of well-researched chapters, each one covering a different field of investigation.

Available at www.amazon.con in paperback and Kindle edition.

Made in the USA
Middletown, DE
15 December 2015